CORTEX AND NETWORK

THE EVOLUTION OF THE INNER MODEL OF THE OUTER WORLD

MICHAEL EVANS M.D.

To Cynthia, Blake, and Alexandria Savanna, usually called Bitzi, who have had to listen to more about tunicates than anyone should.

PREFACE

I began this book to introduce the interesting ideas in neurology to a friend of mine. He turned out to not be as interested in the subject as I was, but I continued to plug away at writing a plausible story of how the brain evolved, how it senses, and thinking through ideas and experiments that have interested me for years.

I ended by having constructed a model of the brain. In trying to explain it, I had produced a rough sketch of the sensory processing side of the nervous system.

The book is about the last six hundred million years.

It is about the evolution over that time of the vertebrate brain's ability to extract and interpret information from the outside world. The major theme is that it does so by making neurological representations of objects in the world, and so models them, and so becomes itself a model of the outside world.

It is in four parts that follow the vertebrate progression from smelling to hearing to seeing to thinking:

> Part I covers pre-vertebrate sea squirts and vertebrate fish. It uses the smell system, the earliest distance sense, as a model of sensory reception, processing, and memory. It discusses the simple models of the world these animals swim through.

> Part II covers amphibians and reptiles and their inner world of sensory representations displayed in a spatial model, the hippocampal map, of the outer world.

> Part III deals with mammals and their sensory representations in auditory and visual cortical maps.

> Part IV deals with humans and abstract conceptual representations of the world. It entertains the notion of mankind having a shared and evolving conceptual model of the entire universe.

I am a clinical neurologist. My subject is not how it works but how it goes wrong. How it works, though, is my motivation for being a neurologist. Ten times a day, with a couple of pins and a tendon hammer, I get to try to figure out a nervous system even though it is usually a broken one. I have used such disease states to illustrate how it works.

I began as an electrical engineer working with signal processing systems. This is what the brain is. It converts sensory inputs into electrical signals and processes them into motor outputs or electrical signals that make muscles contract. Signal processing systems are a focus of the book.

The brain considered as a signal processing system is the brain considered as a model. Simple neural network models of sensing, learning, pattern-recognition, memory formation, and retrieval are developed. Neural networks are interesting in themselves, and this is a brief introduction to them.

The models are not of the of the brain's biological machinery but of its functions, and so functional neural-network models. Since they could be realized on a computer and involve learning, they are functional neural-network machine-learning models, and so simple versions of the models used in artificial intelligence programs.

The book itself could be thought of as a descriptive model. It could add to a reader's mental model of the brain and allow better extraction and interpretation of information from the world of neurology.

A few points on technique:

I have arranged the chapters to sketch the evolution of the brain's ability to extract information from the world. When these abilities developed is not always clear and some are disputed. I have treated the early jawless fish as having pre-wired brains with no ability to learn; or, at least, not mentioned learning until the next chapter. The jawless fish are unlikely to complain, but the odd professor might.

I have emphasized models and minimized neuroanatomy and neurophysiology (function) and ignored neuropathology and neurochemistry. There is so much detail in the leaves and branches of neurology that the trees, let alone the forests, are hard to see.

I have used mathematical equations, but I have not done any mathematics. The equations are another way of representing the models—like looking at the score of a piece of music. Diagrams are another way of representing both equations and models, and I have used them extensively. I have made a case for using mathematical notation to clarify one's thinking, but I have not done anything strenuous.

I have used medical nomenclature (nomen-name, clatura-calling). One of the textbooks of first-year medicine is a medical dictionary to help you figure out the Latin and Greek. Rather than suggest you buy one, I have used parenthetical notes as above.

I did the illustrations. There is one from an old textbook, but everything else is me and a number seven pencil. They could be better but doing them myself allowed me to get them right—at least as I see them.

I have used concepts mostly from the last hundred and forty years, but some from as far back as Democritus and Plato. I have tried to use the original thinker's version of the ideas.

I have omitted disputes about the ideas. A philosophy professor once told me that teaching was a matter of lying and making qualifications. He probably went on to make subtle qualifications. I simply adopted it as my teaching technique. Every tenth statement in this book could be disputed. I have chosen what I think is plausible, made a few qualifications, and skipped the debates.

I have not only lied but prevaricated in its original Latin sense of not plowing within the lines. I have gone beyond the data and speculated. I have, however, flagged my speculations as such.

I have tried to stick to the limited topics of representing and processing. This is an essay. The aim of the essay is to make some of the ideas in neurology interesting enough to read through to the end of the book.

Dunedin, Florida
13 March 2023

CONTENTS

INTRODUCTION: THE BIG PICTURE

Life is sunlight.

Life is a gigantic sun machine. Plants use sunlight and carbon dioxide to manufacture glucose and oxygen. Animals burn glucose and oxygen to make carbon dioxide. Life is sunlight, and the purpose of life, apparently, to manufacture and degrade glucose.

Life is organized sunlight.

Life has run a three-billion-year genetic experiment. As the wheels of the sun machine spun, life evolved entities of increasing complexity through variation of the genomes and selection of the most efficient phenomes. There is nothing new about what we, the phenomes, are made of. Our carbon atoms last time around could have been part of a stick or a sea squirt or a dinosaur. What has evolved is our genome, our DNA blueprint.

Life is informed sunlight.

Life's more highly organized phenomes have highly organized nervous systems. These nervous systems register the information available in the outer world. They perceive it and in so doing form a model of it. The better the model, the more information that can be extracted from the sensory signals available.

Life is not able to fully perceive sunlight or anything else.

Life's perception of the world is limited by the capabilities of its nervous system. The model of the outer world is always incomplete. That limited model is what is known. It is the sensory world the organism inhabits.

Life is able to overcome perceptual limitations with conceptualization.

Life can form abstract concepts of the outer world. These can be improved to the point of near-perfect correspondence with reality. They can never be made perfect, but they are as good as we can get.

PART I

VERTEBRATES AND MODELS OF THE OUTER WORLD

CHAPTER 1

NERVOUS SYSTEMS AND MODELS

I was once a single cell. This is amazing if you stop to think about it. I, the single cell, directed the development of me, the multi-trillion cell entity capable of writing a book about my own cellular development.

To be a little more precise, I consist of 30,000,000,000,000 or thirty trillion cells. Of those cells, 100,000,000,000 or one hundred billion are central nervous system neurons. Each neuron connects to 10,000 others, for a total of 1,000,000,000,000,000 or one thousand trillion for the number of connections or synapses—and all of this began with a single cell.

The single cell doubled and redoubled and became a ball of cells. A pot of cells formed on the inner wall. The pot elongated and folded to become a tube. The tube protruded limb buds. It closed and formed a head.

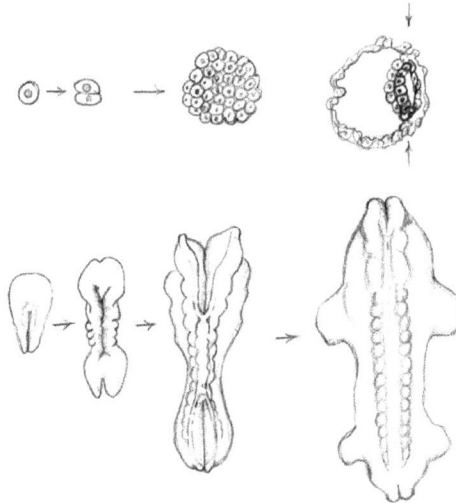

Figure 1 Early Human Embryo Development: The single cell becomes a ball of cells with an inner cavity (upper right arrows) with a surface fold that becomes the early nervous system. It folds further and closes to form a tube flanked by bumpy muscle segments. Limb buds and the beginnings of a head form by four weeks.

The changes in the human fetus were once thought to mimic the stages of evolution. Ontogeny recapitulates phylogeny, or individual development recapitulates species development, in a phrase taught to medical students for the last century.

This was Ernst Haeckel's recapitulation theory and has a certain appeal—we recapitulate evolution—but is wrong. The human fetus never passes through a fish or reptile stage. We share a general body plan and only resemble one another in earliest development. By six weeks we are clearly different in body and brain, as Haeckel's own drawings showed.

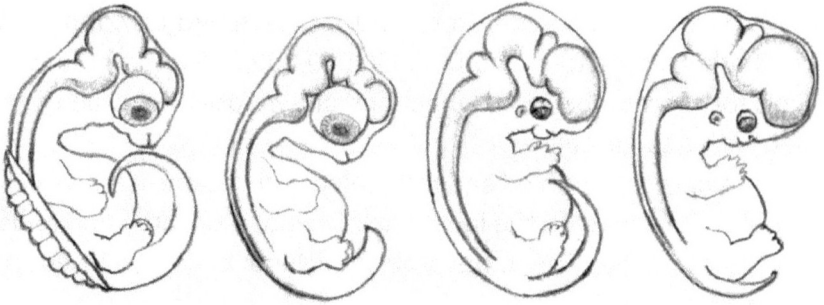

Figure 2 Six-week-old embryos of turtle, chicken, dog, and man with differences in heads and brains, as well as bodies. The folded area is the brainstem. The brain is forward of the folds and the spinal cord below. (After Haeckel, 1868.The lower ends of the spinal cord are faded out where his drawings were imprecise.)

Fetal development can go wrong, but never down the path to a different species. Rabbits do not give birth to fish. Women do not give birth to frogs.

Species Phylogeny

Single celled animals, or protozoa, appeared about four billion years ago. Multi-celled animals, or metazoa, appeared about one billion years ago. The first attempts, poorly organized balls of twenty-five or so cells, are still around. They live only in the intestinal tracts of certain sea creatures.

Pot-like animals were next and included the sponges. They are filter feeders that suck in seawater and strain out food particles.

Tube-like animals were next. Worms abound, but they differ from us in that their nervous systems are in front and wrapped around their intestinal tracts. This makes it difficult to evolve a complex nervous system.

The first identifiable pre-vertebrate is the filter-feeding sea squirt. This unlikely relative has a tadpole-like larva with a cartilaginous rod (notochord), which is the precursor to the vertebral spinal column, and a neural tube, which is the precursor to the vertebrate spinal cord.

It evolved into the first vertebrate, the fish, a form so successful it has dominated the oceans ever since. The fish developed lungs and legs to become an amphibian. The amphibian improved the egg to become a land reptile. The mammal modified the egg to grow internally. Eventually, we came along to write about it all.

Embryonic Development and Neurons

As the cells of the vertebrate embryo multiply, they separate into three types of tissue: skin (ectoderm), gut (endoderm), and connective tissue (mesoderm). The ectoderm (outer tissue) forms an outer tube. It folds in at the bottom to form the gut tube. It folds in at the top to form the neural tube. The mesoderm (middle tissue) forms the notochord and later muscle, cartilage, and bone.

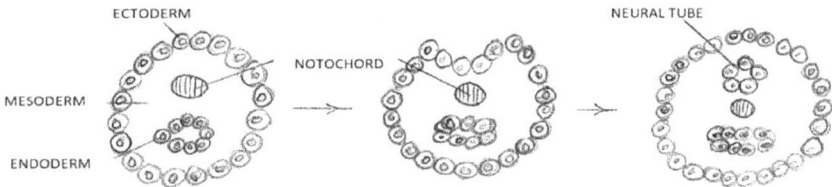

Figure 3 Formation of the Neural Tube: The ectoderm folds in to form the neural tube. The mesoderm is shown without cells for clarity.

The neural tube has nerve cells or neurons. They have cell bodies and those of the large output neurons were thought to look like little pyramids by an early neuroanatomist named Cajal and are still called pyramidal cells. They have input lines called dendrites and output lines called axons. The input dendrites are tissue wires that form a tree-like structure studded with connection points or synapses. The axons are tissue wires that connect to other neurons, muscle cells, and a few glandular cells.

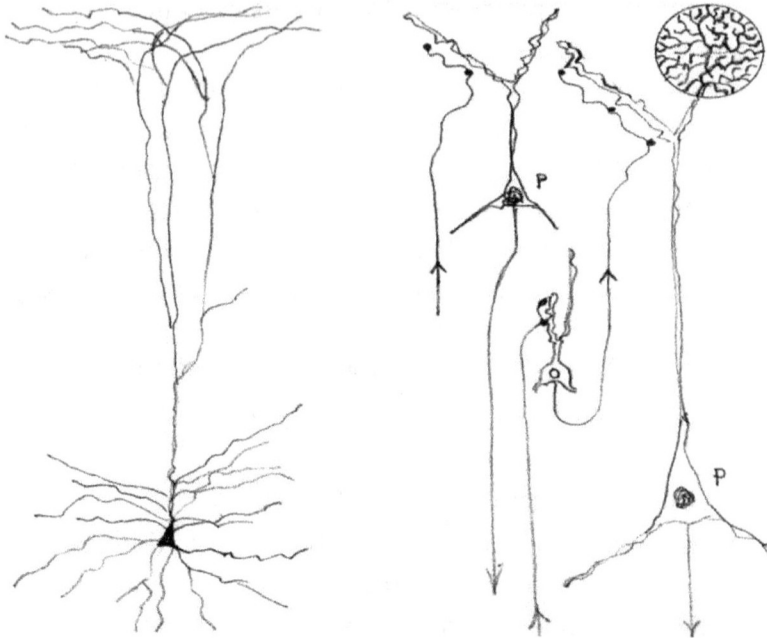

Figure 4 Cerebral Neurons: The drawing to the left is after one of the first pen and ink drawings of a pyramidal output neuron made by Cajal around 1900, with a dark pyramidal cell body and upper and lower dendritic trees. To the right are schematic neurons at higher magnification. The largest is a pyramidal neuron (P) with a circular window showing typical dendritic tree detail. A smaller pyramidal neuron (p) of the kind in the upper neocortical layers is shown above left. Input lines are shown either connecting directly or through a relay cell to black dots that represent synapses. The axons are shown as single down-going lines, although they usually branch extensively.

The neuron is a complicated piece of equipment. It gets streams of electrical impulses from up to 10,000 other neurons. They either excite or inhibit the synapses. The nerve cell body sums and processes all of this and transforms it into an output stream of impulses which it sends down the axon to output synapses.

The neurophysiologist, Charles Sherrington, called the brain an "enchanted loom", a shifting pattern of activated synapses. He also coined the term, synapse, in 1897.

The Brain and Its Electrical Activity

Neurology textbooks have a picture of the brain divided into fifty-two numbered areas by a compulsive, long-dead, Teutonic anatomist, and a picture of a thousand flyspecks said to be neurons. We are not going to go into this much detail. There are seven major areas:

1. The frontal lobe in front of the central fissure directs muscle or motor action in the motor strip near the fissure and more abstract activity such as motor planning in front.

2. The touch or parietal lobe behind the central fissure has a sensory strip that receives touch pressure and joint stretch sensation.

3. The visual or occipital lobe at the back receives light sensation.

4. The auditory or temporal lobe below receives sound sensation.

5. The smell or olfactory bulb below the frontal lobe but connecting to the temporal lobe receives smell sensation.

6. The hippocampus on the inner surface of the temporal lobe maps sensation in space and stores it in memory.

7. The diencephalon (di-across, encephalon-in the head) lies between and connects to the two hemispheres and the brainstem below. It has an area called the thalamus (chamber or entrance room) that relays sensation from the brainstem and spinal cord nerves to the lobes of the brain. It has nerve fiber bundles that connect the two hemispheres.

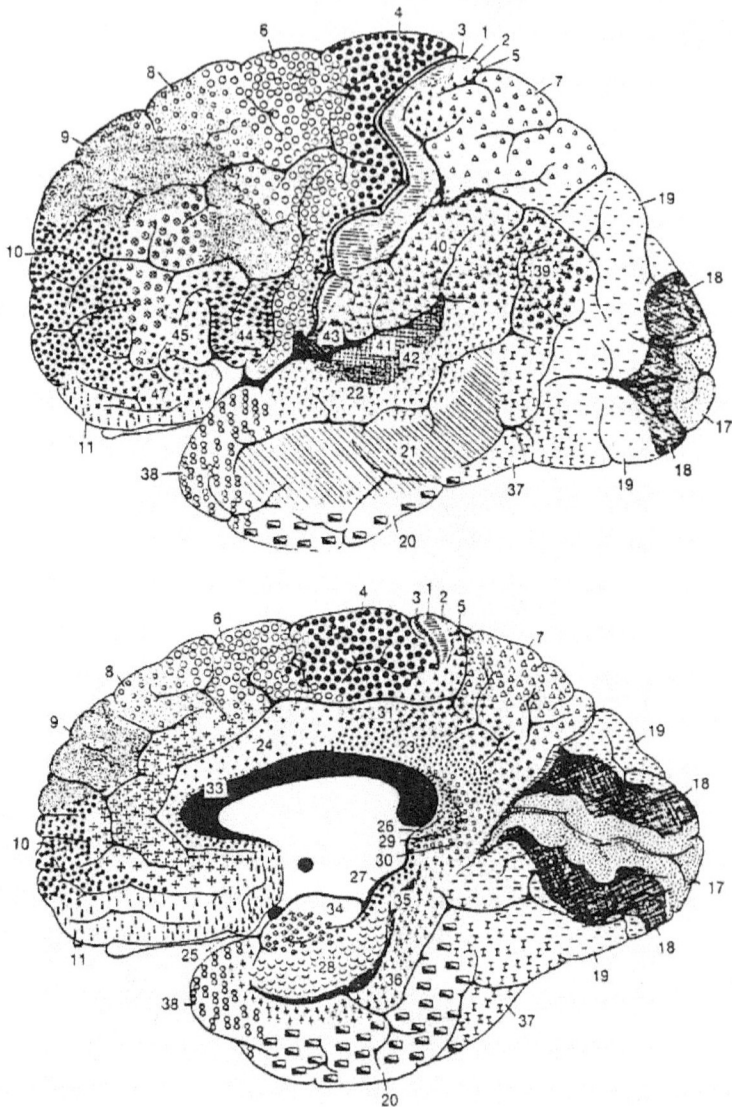

Figure 5 Human brain left outer and right inner surfaces subdivided into fifty-two numbered areas by Korbinian Brodman in 1909. The upper drawing has a central fissure between the frontal motor lobe (4 and forward) and the parietal touch lobe (3 and behind). The occipital visual lobe is at the back (17 and around). The temporal auditory lobe is below (20s, 30s, and 40s). The lower drawing has a white area with a black dot where the diencephalon joins the two hemispheres. The black area (33) is the largest bundle of connecting fibers. The hippocampus (34) is below the black dot. The olfactory bulb (25) is below the frontal lobe.

The anatomical complexity is matched by biochemical and electrical complexity. The neuron uses biochemical pumps to maintain an artificial chemical imbalance with more potassium ions inside and more sodium ions outside. It generates electrical impulses by opening channels in the cell wall to let charged ions move through to create a voltage spike called an action potential. The spike travels down the axon to the output synapses. The neuron then pumps itself back into ionic imbalance and is ready to do it again.

There are three major features:

1. The action potential spike lasts 2 milliseconds. It is started by positively charged sodium ions flowing into the cell and terminated by positively charged potassium ions flowing out.

2. Impulse trains of action potentials at 10 to 100 per second are the messages neurons send to one another.

3. They are also the messages that stimulate muscle cells to contract and glandular cells to secrete hormones in the pituitary gland that hangs below the brain.

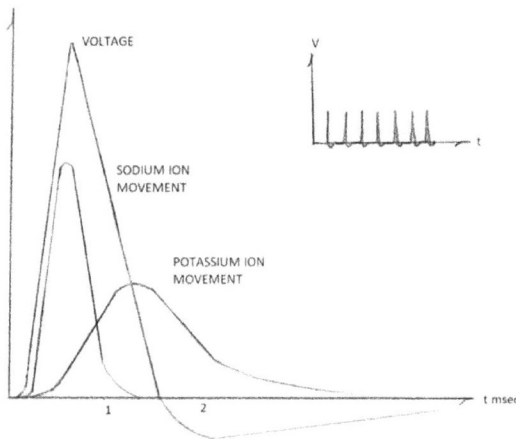

Figure 6 The action potential voltage spike lasts about two milliseconds. The neuron generates it by opening pores in its cell wall to allow sodium ions to flow in, and then stops it by opening other pores to allow potassium ions to flow out. A typical action potential impulse train recording at about ten per second is shown above right. (After Hodgkin and Huxley, 1968.)

Lower voltages cross, and smaller currents flow through the synapses. They, in the dendrites, carry out finer grained computations than does the neuron itself. We are going to simplify and consider only whole-neuron processing.

Model Neurons

We are going to further simplify by using model neurons, also called formal neurons. These neurons are simple indeed—they add their inputs.

The simplest of the simple is an engineer's device, an on-off switch. It has many inputs and one output. If enough of the input lines are on, it turns on and sends an on-signal to the next neuron; if not, it stays off.

It was first used for neuron modelling in the 1940's, not by engineers, but by Warren McCullough and Walter Pitts, who were philosopher-psychologist-physician and eccentric-genius-hanger-on, respectively. They used it to make somewhat implausible brain models. McCullough was a Quaker with an old-testament prose style and their paper was called, "A Logical Calculus of the Ideas Immanent in Nervous System Activity".

To an electrical engineer, this is a two-state or binary machine, and its output state can be written as a binary bit (one or zero). The neuron adds its four inputs (of one or zero) and, if the sum is greater than the threshold value (three in the example), the neuron decision box sends a signal (a one) down its output line.

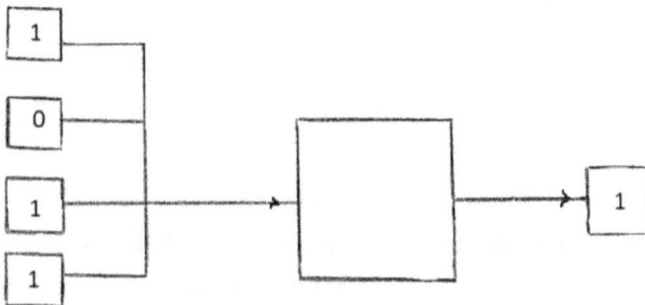

Figure 7 Binary Threshold-determined On-off Switch. It sums its inputs. If the sum is more than its threshold value, its decision box signals an output of 1; if less, zero.

It runs from a clock and with each tick it does its addition. All computers in use today have a two-state architecture and a clock pulse that makes them essentially McCullough-Pitts neural networks. Their decision boxes are the rules of binary arithmetic.

The brain, however, does not work with bits and its neurons do not have decision boxes. The model evolved into the summation model neuron of the 1970's which operates not with pulses but pulse firing rates, and with decimals. It gets inputs from model neurons (numbered from 1 to n), and adds their firing rates (f1, f2,...,fn) which could be one, zero or some other number, and generates its own output firing rate (g),

$$g = f1 + f2 + \dots + fn.$$

The McCullouch-Pitts model neuron is binary: on or off, one or zero, black or white. The summation neuron can vary over a range of values, and represent black, white, or shades of grey. If, however, the input values are ones and zeros as in most of our examples, it is reduced to working like a McCullouch-Pitts formal neuron.

The input neurons can vary in how strongly they connect. This is represented mathematically by connection strength numbers, called connection functions (a1, a2,...,an), and the summation neuron equation then becomes

$$g = a1.f1 + a2.f2 + \dots + an.fn.$$

The connection functions are the mathematical versions of synapses. They determine how input signals affect the neuron and so how it processes them. If the connection function is large, the input affects the receiving neuron a great deal; if small, a little; if zero, not at all.

The mathematical symbols are the standard plus operator (+) for addition and dot operator (.) for multiplication. At times in this book, they will represent more complicated addition and multiplication operators.

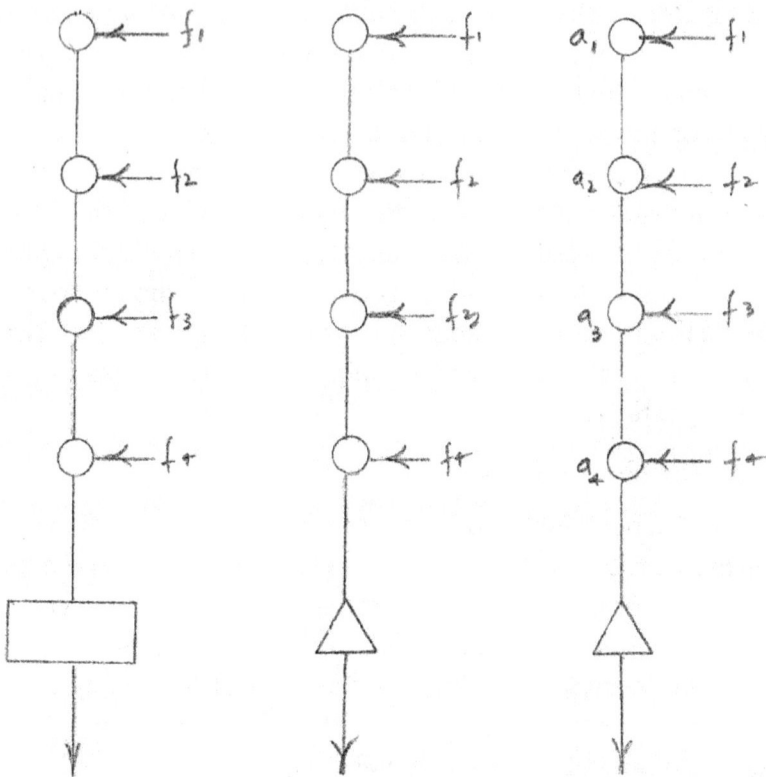

Figure 8 Model or Formal Neurons: The McCullough-Pitts model neuron on the left sums its inputs f1 through f4, and the rectangular decision box sends on a 1 if the sum is equal to or greater than its threshold value. The summation neuron in the center sends on the sum, g=f1+f2+f3+f4. The connection function summation neuron on the right sends on the sum, g=a1.f1+a2.f2+a3.f3+a4.f4.

    The model neuron can represent the three types of real neuron: the motor output neuron, the sensory input neuron, and the interneuron that acts between the two. It can represent a single synaptic input or many.

    The synapse can be represented as a circle with a cutout shape that a stimulus object can fit into. The stimulus object can be a sensory input for a sensory neuron or a neurotransmitter input for an interneuron or output neuron.

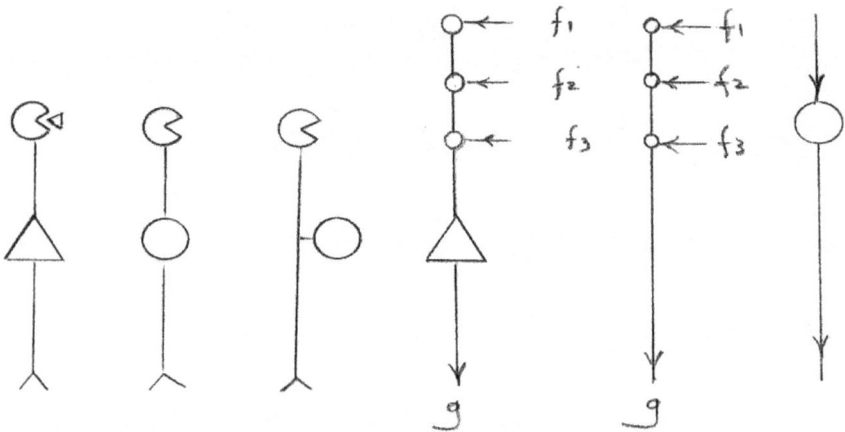

Figure 9 Model Neuron Representations: At left is a schematic pyramidal or output neuron with a dendritic tree consisting of one input synapse diagrammed as a circle with a cutout that a triangle-shaped stimulus object can fit into. Next is a schematic interneuron. Next is a schematic sensory neuron where the input area is a sensory receptor, and a continuous dendrite-axon goes to the output synapse, and the cell body in a real cell can be separated from it as shown or like that of the interneuron. A longstanding convention is that the axon synapse terminations are reversed arrowheads. We will usually be using simpler representations, as in the fourth model neuron, with multiple input synapses diagramed as small circles with arrow inputs and with arrow outputs going in the out-direction. The cell body symbol can be left out without confusion as in the next diagram. The input can terminate on the cell body symbol without confusion as in the last diagram.

Model Neural Networks and Algebraic Descriptions

The summation model neuron is a reasonable first approximation to a real neuron, and all we will need to make models of the nervous system.

These will be neural network models with multiple inputs (f1, f2, or more) and multiple output neurons (g1, g2, or more). When things get this complicated, the numbering schemes do as well, and the convention is that the connection functions get double-barreled numbers like (a21)

g1=a11.f1+a21.f2

and the equation network diagram looks like this.

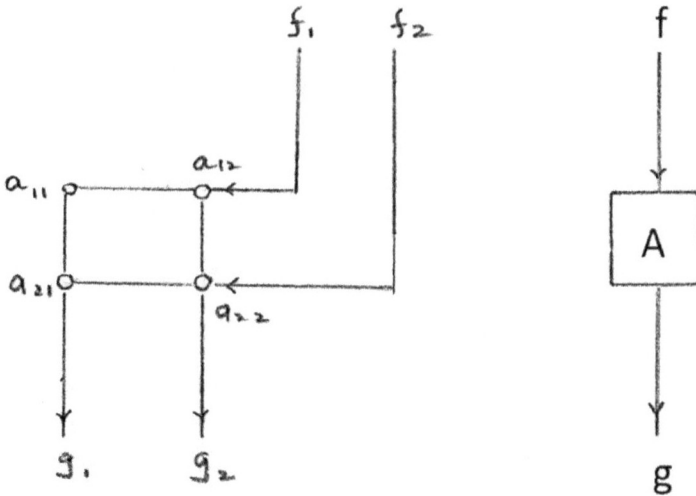

Fig 10 A neural network and its shorthand description with inputs terminating on the network symbol (A).

The shorthand algebraic description of such a network is

$g = A. f$

where g and f now stand for the sets of output and input lines, and A stands for the set of connection functions, and the dot operator is now a more complicated operator that multiplies the f's by the correct subsets of a's to produce the g's.

A lot of information is packed into this shorthand representation. We will use it when we discuss networks and not struggle with double subscripts or other details.

The input to the neural network can be thought of as a simple pattern like black-white and coded as (1,0), and its output a changed pattern such as (1,1) or black-black. This is what would happen in the example if (a11) and (a12) were both 1.

Neural networks can have multiple layers or, to put it another way, one network can feed another.

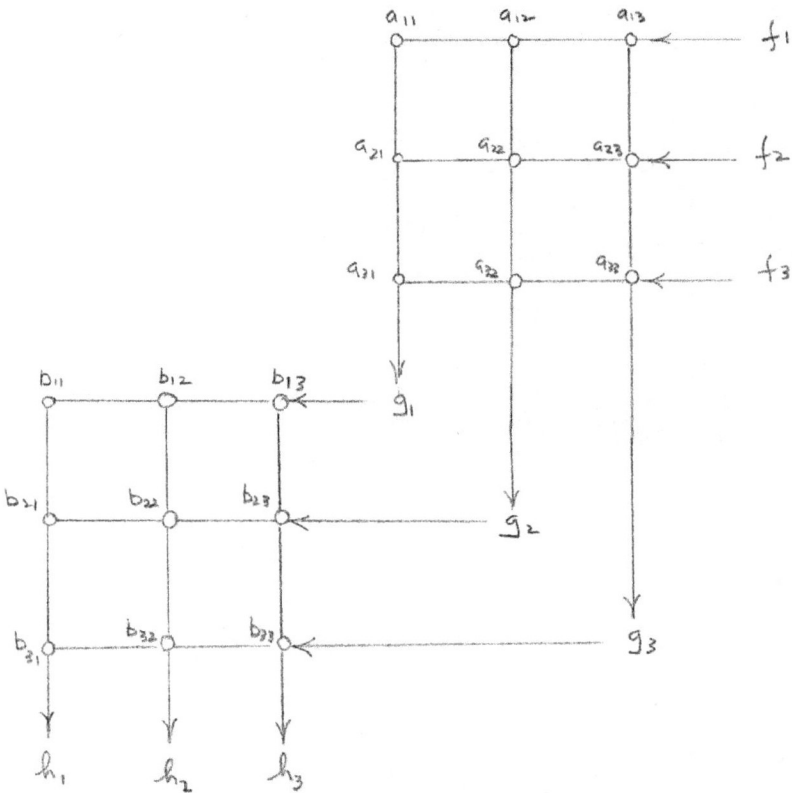

Figure 11 Multi-layer Neural Network Diagram: The upper right network has inputs (f1, f2, f3) and output neurons (g1, g2, g3). The g neurons go on to another neural network which transforms g inputs into h outputs.

Multi-layered networks process their inputs in stages

$$g = A.\ f$$

and

$$h = B.\ g.$$

and each stage is a different transformation of the input which can be sent on to another network for further processing or to an action network.

More layers mean more capabilities. Loosely speaking, the more layers in a neural network, the more it can do. In artificial intelligence applications, there are many layers of many model neurons. The layers between input and output are called hidden layers. Adding a third layer to our example would make the B-layer a hidden one.

The brain works like this, although not quite this simply. It processes inputs sequentially with sensory signals transformed and sent on to further sensory processing stages, motor directing stages, or memory stages. It also processes inputs in parallel pathways and can combine them.

Model Neuron Learning

A summation neuron model can learn. It does so by changing its connection function (a) by an amount (Δa) to a new value (a +Δa) so that the pre-learning output

$$g= a. f$$

becomes the after-learning output

$$g= (a+Δa). f$$

and the neuron has learned to change its response to the same input.

A connection function pattern change (ΔA) contains many Δa changes, so that

$$g= (A+ΔA). f$$

and this changes the transformation of the input pattern. If (a11) were changed to 8 and (a12) to 4, then the [1, 0] input would be transformed into [8, 4]. We will discuss learning rules and how this happens later.

One of the artificial intelligence techniques of machine learning is learning to recognize patterns with neural networks. When the learning takes place in the hidden interior layers, it is called deep learning. We will discuss pattern recognition later.

Auto-associating and Feature-detecting Networks

There are two networks that we will encounter in later chapters:

The first is the auto-association network which sends its outputs back to its input lines. It has feedback loops.

Real cerebral cortex does this. Cortical neurons send axon branches, called recurrent collaterals, back to their own dendritic trees to make feedback loops. They can also go through interneurons to form longer loops. Such loops are good for memory storage and recovery. They are also good at retrieving a complete memory from a partial input. The first output can trigger the full memory output on the second pass through.

The second is the feature-detecting network. Real sensory cortex has focal spots that register specific inputs like lines or faces. They are neurons or groups of neurons that are activated when certain features are detected and are called feature detectors or hot spots.

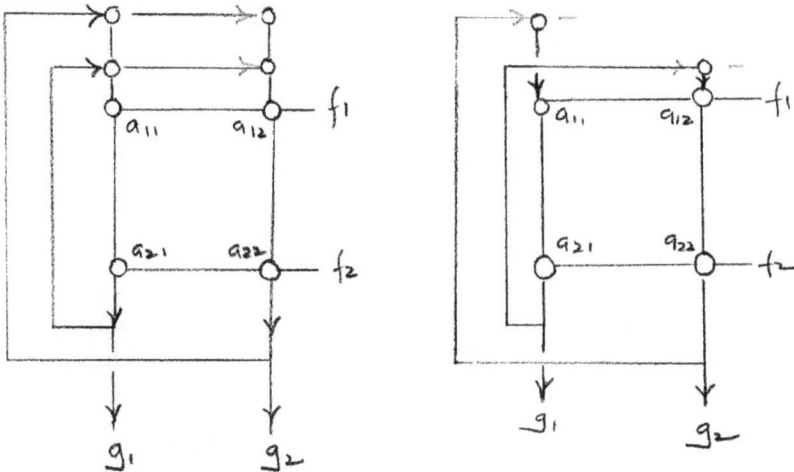

Figure 12 An auto-association or recurrent loop network is on the left. Each neuron sends a feedback branch to itself and every other neuron. A self-organizing feature-detecting network is on the right: If g1 sends an inhibiting signal to g2 connections then g1 would become a small hotspot. The lower feedback connections are not shown for clarity.

Recurrent networks with learning can organize themselves to become feature-detectors. Tuevo Kohonen in the late 1970's found that if model neurons sent inhibiting feedback signals to surrounding network neurons, the most activated neurons would tend to become hot spots against a flat background of inhibited neurons. In our example, the most active neuron could inhibit the other and become a one-neuron hot spot.

This is called a self-organizing feature-detecting network. We are not going to work with such networks, but we are going to discuss experiments done with them.

Real Sensory Neurons and Receptors and Reflex Responses

A psychologist in a bad mood once made the dismissive statement that, "...in common with single cellular organisms, all we can do is discriminate among a paltry few of the potential stimuli in our environment and then contract, relax, or secrete."

Discriminating is the business of the nervous system. Contracting and relaxing, the business of the muscles. Secreting (of hormones) is done by glandular cells, and those in the pituitary gland just below the brain are controlled by brain neurons.

The cells that discriminate among the paltry few environmental stimuli are sensory neurons. They have receptors for light, sound, touch, smell, and taste.

The receptor is a fundamental concept in neurology and in medicine. It originated around 1900 in the work of John Langley and Paul Ehrlich who found that drugs acted on cells by binding to receptors embedded in their cell walls.

The concept was extended to the action of neurotransmitter chemicals. At the output synapse, the electrical signal causes neurotransmitter molecules to be secreted into the synaptic cleft where they act on the receptors in the dendritic trees of the receiving downstream neurons. The electrical message becomes a chemical message. Drugs can act on these receptors also.

The concept was extended to the receptors of sensory neurons which are excited by molecules for smell and taste, and by the physical stimuli of light, sound, and pressure for vision, hearing, and touch.

Receptors are transducers: devices that transform one kind of signal into another. Sensory receptors transduce their input stimuli into streams of electrical impulses like telegraph signals. The impulse trains go to the output synapses where the electrical signals are transduced into chemical signals. The dendritic receptors of receiving cells then transduce the chemical signals back into electrical signals.

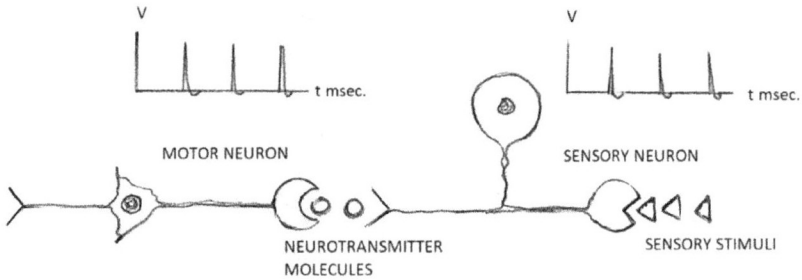

Figure 13 Sensory Neuron and Dendritic Synaptic Transduction: Sensory stimuli (triangular objects) insert themselves into the receptor that transduces them into an electrical signal. The sensory nerve generates an impulse train and conducts it to its output synapse where it is transduced into neurotransmitter molecules (circular objects) that are released to insert themselves into the synaptic receptors of the receiving neuron, here a motor neuron, which transduce it back into an electrical impulse train.

If the next cell were a muscle cell, it would respond by contracting. This would be the simplest possible reflex. If the next cell were a neuron, the message would move on.

Our simplest reflex uses two neurons: After the sensory neuron comes a motor neuron that stimulates the muscle cell. Our knee jerk reflex results from a tap to the knee that stretches a tendon, that excites the stretch receptor of its sensory neuron, that excites a spinal cord motor neuron, that excites the thigh muscle of the tendon, that contracts and jerks the lower leg.

This is the oldest test in neurology. The philosopher, Rene Descartes, first described the reflex, although not this one, in 1662.

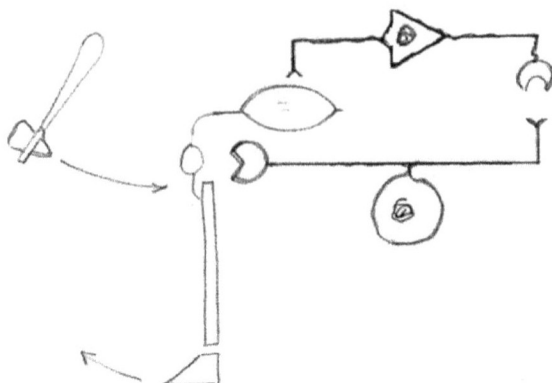

Figure 14 Knee Jerk Reflex: A tendon hammer blow to the tendon connected to the kneecap mechanically excites the stretch receptor of the sensory neuron, which stimulates the spinal cord motor neuron, which stimulates the thigh muscle to contract and jerk the leg.

Even the worm has motor reflexes like this. It responds to a touch by coiling into a ball. The difference between the worm and ourselves is that we have more levels of nervous system: layer after layer of more sophisticated reflexes adding complexity to our responses to the environment. It has been argued, by behavioral psychologists, that this includes brain function—highly processed responses with long delays.

The upper levels over-ride the lower ones, but the older responses are still there, as anyone who has ever stepped on a tack or visited a neurologist's office knows.

Nervous System Design

The simplest nervous system design would be a single sensory neuron. What would be the point? The response would have nowhere to go. A message without a recipient is pointless. Knowledge without the possibility of action is useless. (The jellyfish gets around this by having combination sensation-action cells that not only register a sensation but also fire a dart.)

The next design would be a two-cell nervous system. A sensory neuron stimulating a muscle cell.

The next would be a three-cell system with a sensory cell stimulating a motor neuron in turn stimulating a muscle cell: the knee jerk. This is called a one-synapse or monosynaptic reflex.

The next step is to put another neuron between the sensory neuron and the motor neuron. The in-between-neuron is called an interneuron.

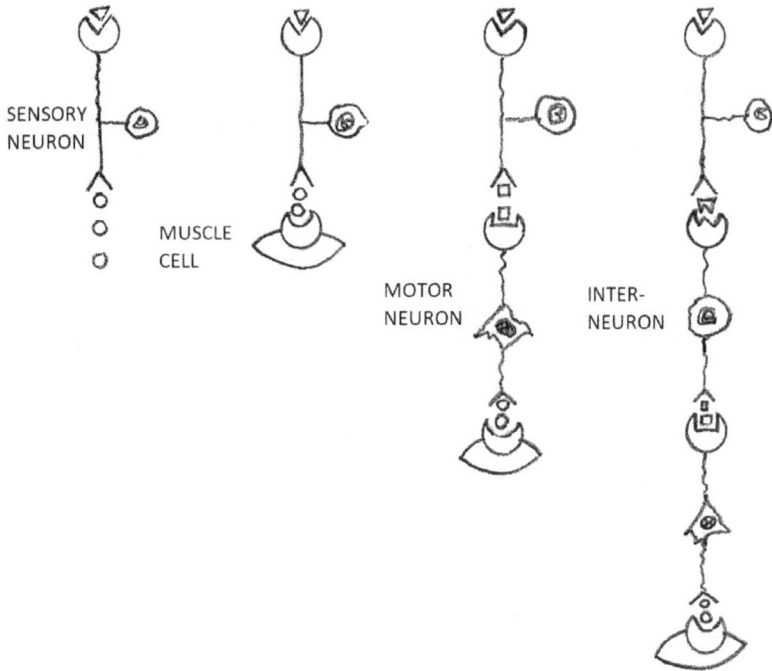

Figure 15 Nervous System Design: To the left, an isolated sensory neuron sends neurotransmitter molecules into empty space. Next, a sensory neuron acts on a muscle cell receptor. Next, a sensory neuron acts on a motor neuron which acts on a muscle cell. Last, a sensory neuron acts on an interneuron which acts on a motor neuron.

This is a major step. The interneuron allows modification of the incoming sensory signal before it influences motor output—of information processing.

The simplest interneuron system would be one or more inter-neurons that would allow some elaboration of a reflex response. We have polysynaptic spinal cord reflexes of this sort, like the withdrawal reflex to stepping on a tack with movement at toes, ankle, knee, and hip. The worm has them too, although not that particular reflex.

The most complex interneuron system is the human brain. It is a set of interneurons between sensory input and motor output.

The next step in theoretical nervous system design is the bilateral plan. The bilateral body with a bilateral nervous system is a design that can be improved upon. The random plan and the circular plan lack a clear direction of neurological effort. The worm is bilateral, as we are.

A bilateral nervous system can be segmented so that each nervous system segment acts on its local muscle and body segment. Our spinal cord is like this. The worm's is also.

The bilateral plan can act in either direction, and the next step is deciding on a preferred direction and putting the head there with sensors to detect what is out in front. It seems best to put the head on the mouth end—at least we have not found any organisms built the other way. The worm has a head too.

The upper part of this nervous system is mostly concerned with mouth and head activities. The upper continuation of the spinal cord called the brainstem is like this. The worm does not have a brainstem although it does have more nerve tissue in its head than its tail.

The next step is to put distance sensors in the head. The primitive fish did this and, in the process, developed the brain, the front end of which was a chemical sensing system for smell. At this point we leave the worm behind and join the vertebrates.

The distance sensor is a big step: It tells an animal what is out in front before it gets there. Richard Gregory pointed out that this freed the animal from "the tyranny of the reflex" and was the precursor to intelligence.

The human brain is the fish brain evolved, with more distance sensing and more processing.

Clinical Neurology

The upper end of the human neural tube closes and balloons out on each side to form the cerebral hemispheres. This can go wrong and the tube not close, the hemispheres not form, and the malformed head remain open. This is called anencephaly (an-no[thing], en-in, cephalos-head), and fortunately is a very rare developmental malformation.

Human or clinical neurology makes four basic distinctions:

Between the autonomic system that controls the body organs and deals with the inside world, and the somatic system that controls the body muscles and deals with the outside world.

Between the nerves in the peripheral nervous system of the body and those in the central nervous system.

Between the three central nervous system sub-divisions of spinal cord which controls the arms and legs, brainstem which controls the face, and brain.

Between the three kinds of brain areas: Those that process sensory inputs. Those that direct motor outputs. Those that work between the two in what are called association areas where we consider sensory inputs and plan motor outputs—and think.

Touch sensation comes into the human nervous system through a sensory neuron in a peripheral nerve that enters the spinal cord and relays to a second and central neuron. It goes up the cord to relay on a third neuron in the thalamus, which directs it to the neurons of the touch brain.

The brain is modular; in addition to the touch area, it has separate receiving areas for smell, taste, hearing, and vision. Their outputs are sent to higher sensory processing areas and to frontal areas for motor control.

A neurologist must do a two-part diagnosis, first localizing the symptoms to some part of the nervous system, and then working out the cause. A numb hand could localize to nerve, spinal cord, or brain. The cause could be nerve compression, cord compression, or brain blood vessel clot.

Since one side of the brain connects to the nerves of the other side, damage to the touch area causes numbness of the opposite side of the face and/or body. A small amount of damage could affect just a hand or finger.

Damage to the spinal cord causes numbness on both sides below the level of damage. Disease of the peripheral nerves, called peripheral neuropathy (nerve sickness), causes numbness too, usually first in the longest and most vulnerable nerves causing the feet and hands to go numb.

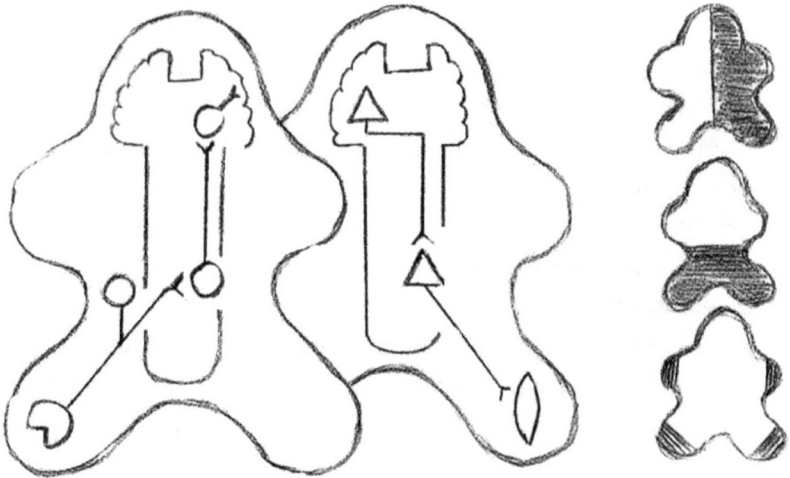

Figure 16 Basic Gingerbread Man Neurology: A touch signal from the receptor of the sensory peripheral neuron crosses to an interneuron on the other side of the spinal cord, to an interneuron in the thalamic relay center, to the sensory cortex of the brain. A motor cortex output neuron sends a signal down a long axon that crosses in the brainstem to the opposite side of the spinal cord to stimulate a motor neuron in the cord which stimulates a muscle cell. The smaller figures show the sensory deficits of brain, spinal cord, and peripheral nerve damage.

Motor control requires only two neurons: A motor cortex neuron sends a signal down a long axon that crosses to the opposite side of the brainstem to stimulate a lower motor neuron in the brainstem or spinal cord, which in turn stimulates a muscle cell.

Damage to cortical motor neurons causes opposite side weakness; to cord, weakness below the cord level; to nerve, focal muscle weakness. Leaning too much on the wrist or the palm of the hand can damage the median nerve to the thumb, causing carpal (wrist) tunnel syndrome and inability to wiggle the thumb.

A blood clot blocking a large blood vessel in one hemisphere of the brain can cause extensive damage with both numbness and weakness of the opposite side of the body. Centuries ago, this was thought to happen when a witch stroked that side. They are still called strokes, although the preferred term is cerebrovascular accident. Strokes of different sizes in different areas have shown us how various parts of the brain work—or rather stop working. They are experiments of nature. They are the neurologist's probe of brain function.

Clinical neurology is a little more complicated than this discussion and these gingerbread man diagrams would suggest.

Nervous System Evolution

Brains get bigger through evolutionary time. Brains have to do with intelligence. Karl Lashley pointed out that the only known physical correlate of biological intelligence is brain size. It is also the only physical measurement you can make of the brain of an extinct animal, a cast of the inside of the skull or endocast.

In comparing brain sizes, you must compensate for body size. A neuron can control at most a thousand muscle cells, so larger bodies with more muscles need more nerve cells to control them.

Harry Jerison compared brains by graphing brain weight against body weight for hundreds of animals' brains and extinct animals' endocasts. He found that they fell into three groups: The fish, amphibians, old reptiles and modern reptiles fell in the lowest; the early mammals, next; and the modern mammals, the highest.

The easiest way to appreciate his work is to normalize all animals to a standard size, say that of the 70-kilogram or 154-pound average-man of clinical medicine, and compare average brain sizes. This corresponds to following the 70-kilogram line up his many double-logarithm charts.

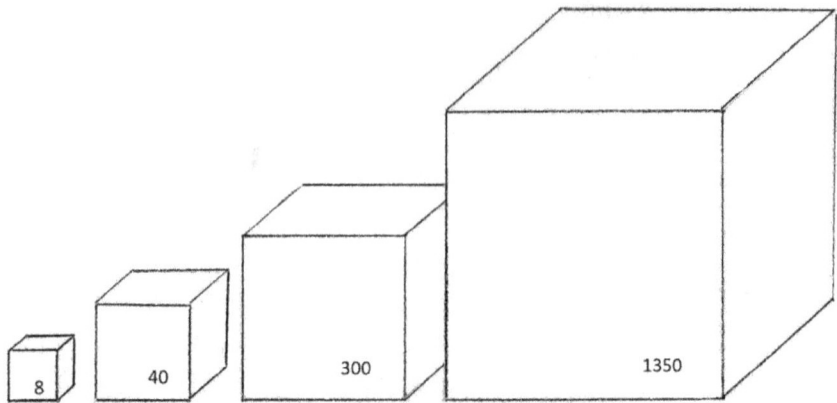

Figure 17 Normalized brain weights in grams for 70- kilogram jawed fish and frog and reptile (all 8), early mammal (40), later mammal (300), and human (1350).

We can then see the brain size increases:

The normalized 70-kilogram jawed fish has a brain weight of 8 grams. (The early jawless fish brain was probably even smaller.)

Fish to amphibian to early reptile development involved no change in brain size. The move onto land required lungs and feet—not a bigger brain.

Modern reptiles have brains that are no larger than primitive ones. Reptiles do not exploit the brain.

Early mammals did exploit it, mostly for hearing. Theirs were five times larger than the reptiles at 40 grams.

The more recent vision-driven mammals of the last 50 million years exploited it further with enlargement to 300 grams. The primates were on the high side at 450 grams.

Humans went even further with a two-and-a-half-times-increase over the primate brain to 1350 grams. The 70-kilogram human has more than thirty times the amount of nervous system needed by an early mammal and almost two hundred times that needed by a fish, frog, or reptile.

Bigger is better. Sophisticated design is all very well, say engineers, but you can just make a bigger hammer.

Why did brains get bigger?

Not all did. As vertebrates evolved, the newer species tended to have some with larger brains, and so it was variation in brain size rather than size itself that increased. The smaller-brained continued quite successfully in the old small-brained ways and the larger moved into more demanding lifestyles.

Brain size seems to be not a target of evolutionary development but a Darwinian characteristic like webbed feet that allows an animal to be successful in a particular niche. One ecological niche that would reward brain size is a predator-prey one. In such a competitive relationship, an intelligence arms race could develop with each gain on one side forcing a gain on the other.

What can bigger brains do?

Brain weights do not reveal changes in the way the brain works, but comparisons of different species do. They reveal the emergence of more sophisticated functions:

Simple animals respond to stimulation of their receptors with motor reflexes. They are stimulus-response mechanisms. The receptor, like the reflex, is tyrannical. It can detect only what its structure allows. Its responses are fixed.

A group of receptors, however, can recognize a pattern. If it has memory, it can learn a pattern. If it has a lot of memory like the human brain, it can learn many complicated patterns and equally complicated responses to them. This is perceptual recognition, and it allows an organism to overcome the tyranny of the receptor.

At a more abstract level, a pattern can trigger a concept. Conceptual recognition allows the overcoming of the tyranny of the percept. Action can occur after prolonged and sometimes abstract thought. The brain can become a mind.

Engineers say sophisticated design can work too.

Emergent Functions

Higher brain functions such as thinking and speaking, and even walking, emerge in brains that previously did not have such abilities. Emergence smacks of discredited theories like spontaneous generation, and seems somewhat murky and mystical, but is just nature's way of taking whatever is available and using it for a new job: If fish start squirming across mud flats, those with the thickest lower fins will do best, and will evolve to make walking with their emergent legs better.

Nervous system emergence has taken place six times in our line of development. (What constitutes an emergence is a matter of opinion. These are mine.)

The first was 500 million years ago when the smell detection system of a primitive fish started retaining traces of previous smells and developed a memory-using processing system or brain.

The second was 165 million years ago when the dinosaurs took over the earth and relegated early mammals to the marginal role of small night predators. Hunting at night with only occasional sounds as signals led to brain representations of prey in an internal model of the outer world.

The third was 80 thousand years ago when hominids began to bury their dead and make cave paintings. This change in behavior took place long after brain size had stopped increasing. There was no change in size or structure, but there was a change in use. This seems to indicate the emergence of mind.

A fourth took place in that mind 40 thousand years ago when men began to use complex syntactic speech.

A fifth emergence was 2.5 thousand years ago when language became sophisticated enough to allow abstract and critical thought, both aided by writing. A refinement, 200 years ago, was the scientific method.

A sixth emergence was 100 years ago when the neuroanatomist, Santiago Ramon y Cajal, first saw the individual neurons that did the work of the brain and showed that it was not just amorphous gelatin. The use of the mind to understand the brain is quite recent.

CHAPTER 2

PREVERTEBRATE SEA SQUIRTS AND RECEPTORS AND VON UEXKULL'S SENSED WORLD

Whenever I want to contemplate a 600-million-year-old nervous system, I wander down to the end of the dock and look at the pilings. Where I live, in Florida, there are lots of pre-vertebrate nervous systems stuck to the wood. They are not very impressive.

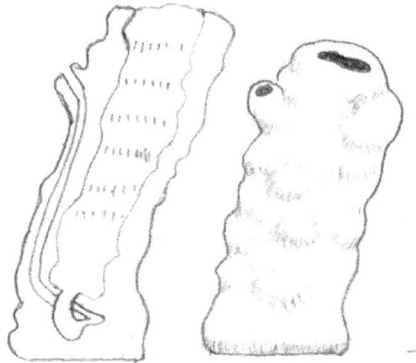

Figure 18 The tunicate, branchiostoma floridae, of the Gulf of Mexico: About 4 inches long. The cross section shows the input siphon opening into a large pharynx leading to a stomach and intestine and then output siphon.

The nervous systems are housed in thumb-sized gray bumps with two holes at the top. Occasionally, one of the holes emits a squirt of water. That is how they got their common name, sea squirt, and that is all they seem to do. Another name is tunicate because they have thick tunic coverings. If you cut one open, you find almost nothing but gastrointestinal tract. They are stand-alone guts. They suck in water and squirt it out. On the way out, they strain out the food particles. They are sessile filter feeders. There are probably a few where you work.

We are going to discuss the tunicate's limited ability to sense the world around it.

Tunicates

Tunicates appeared around 600 million years ago (MYA). They are not simple pots. They have sophisticated hearts and guts and are far along the evolutionary road from the simpler sponges they resemble.

They have gill slits in a large throat or pharynx leading to a stomach. They do not use them to breathe as fish do but to strain out food particles in the water.

These examples of pre-vertebrate nervous systems have almost no nervous system. They have a few neurons around the mouth for mouth-closure reflexes. They have a few pacer neurons that cause periodic contractions of the gut and the squirts. Their entire nervous system is a primitive form of the autonomic nervous system that controls our hearts, guts, and sphincters. There is neither a sensory system that can register the outer world nor a motor system to act on it. There is nothing much at all.

These minimalist adult nervous systems, however, are not the entire story. These somewhat boring animals began as larva, and that nervous system is interesting: It has a neural tube, which is the evolutionary precursor of the vertebrate spinal cord, and which sits above a cartilaginous notochord, which is the precursor of the vertebral column. This is their defining characteristic and gives the general name to this group of animals, chordates.

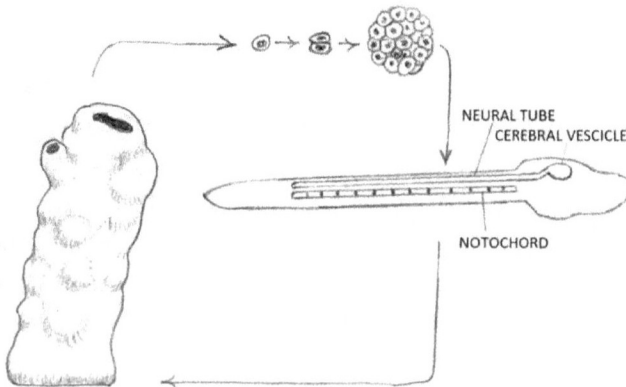

Figure 19 Tunicate Life Cycle: From adult to offspring single cell to larva with notochord, neural tube, and cerebral vesicle containing sensory organs.

Tunicate Embryology and Larva

The tunicate begins as a single cell and multiplies to form a ball of cells. The ball indents to form a pot, which closes to form a gut and elongates to form a worm shape with open ends. This double gut-opening allows flow-through and is an improvement on eating and excreting through the same opening as sponges do. Only double-enders evolve into advanced animals.

Along the top of the gut, a line of cells forms the notochord plate and then rod. The skin cells over the notochord fold in to form the neural plate and then tube.

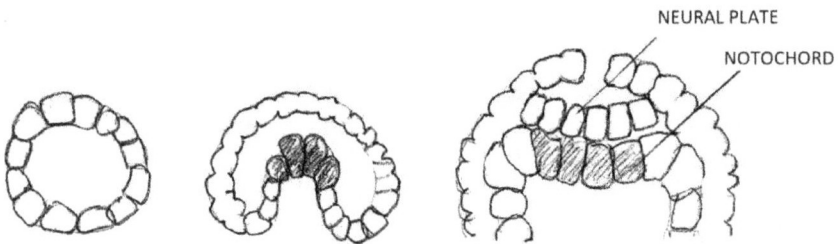

Figure 20 Tunicate Embryological Development: The ball of cells at left folds in to form a pot that closes below to become the gut. The dark gut cells separate and form the notochord. The skin cells above the notochord fold in and form the neural plate. Later the notochord folds into a rod, the neural plate folds into a tube, and the skin cells close over the top.

After two days, the completed larval tunicate looks like a poorly drawn tadpole. The blobby front end contains the adult gut, but it is turned off. The tail has the forty cartilaginous cells of the notochord, three columns of eighteen muscle cells, and the neural tube with motor neurons to control the muscle cells. The front of the tube balloons into the cerebral vesicle (blister in Latin) with a light-detecting eyespot and a gravity-detecting organ called a statocyst (stato-standing, cyst-bladder in Greek).

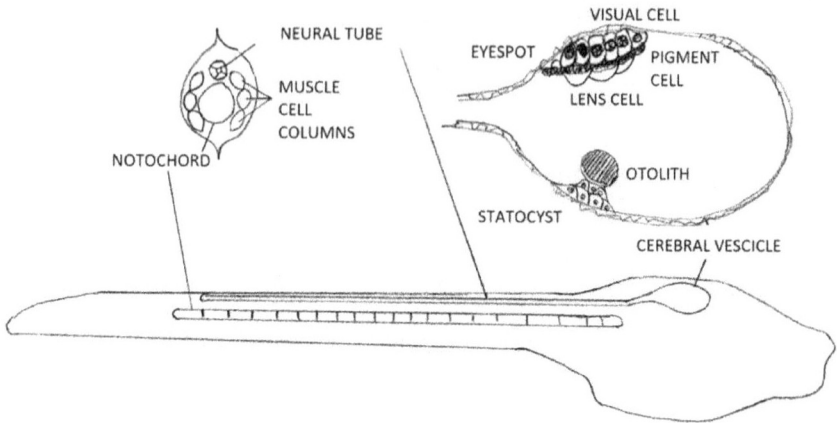

Figure 21 Larval tunicate with the notochord and neural tube and cerebral vesicle with the eyespot and statocyst. The tail cross section shows the notochord and neural tube and the three muscle cell columns on each side.

The behavior of this animal is simplicity itself. After two days of development, it becomes a surface seeker and swims toward light and away from gravity. When it reaches the water surface, it drifts for two days. It then becomes a bottom seeker, and swims away from light and toward gravitational pull. When it reaches the bottom, it attaches to whatever is around, a rock or a dock or a sailboat. It turns on its filter feeding gut and never moves again.

Once it has attached, it no longer needs a nervous system. A hormonal signal is secreted and the animal metamorphoses. The nervous system and tail degenerate and are re-absorbed into the now simplified animal. It retains a few nerves to control its gut.

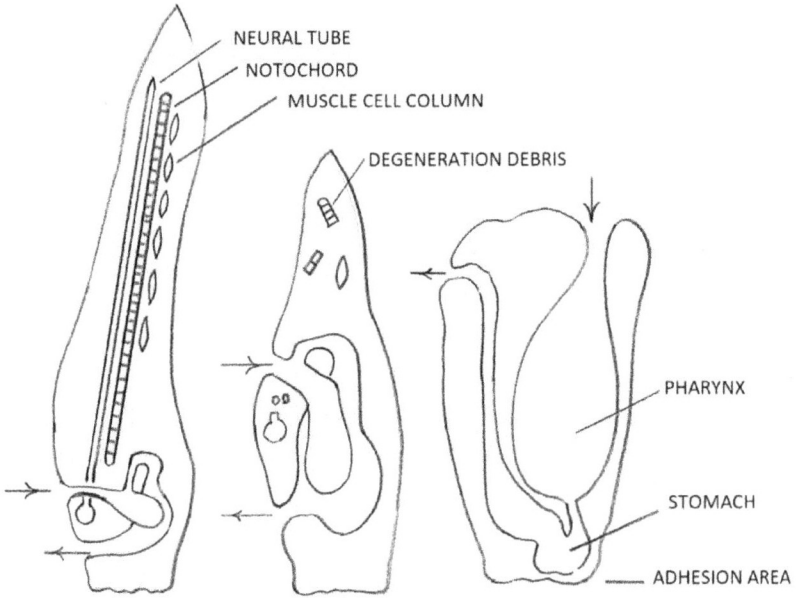

Figure 22 Tunicate metamorphosis and nervous system re-absorption: The front-end adheres, the tail degenerates, is re-absorbed, and the mouth ends up on top.

Nervous System Block Diagram Model

What kind of a nervous system does the larval tunicate need?

It needs a light detector, a gravity detector, an action or motor system, and an interneuron processing system to link sensory input to motor output. The block diagram has four boxes, one for each function, and the muscle cell columns.

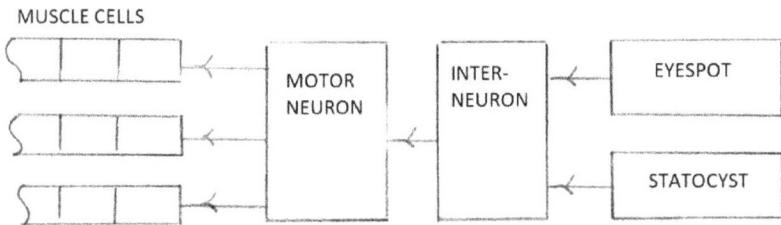

Figure 23 Tunicate Nervous System Block Diagram: The sensory organs activate the interneuron network that activates the motor neurons that activate the muscle cells.

The light detector is an eyespot: a cluster of cells with light receptors. They contain photoreceptor molecules that undergo chemical change and generate electrical impulses when struck by photons. These cells are like those in our own eyes, but our eye structure is more complicated and more useful—all the eyespot can do is register the presence of photons.

The gravity detector is a statocyst: cells with a projecting tip capped with chalk called an otolith (ear-rock). Under the influence of gravity, the tip bends and stimulates the cells to generate electrical impulses. These are like the neurons in the balance areas of our inner ears. We have more such cells, and a more complicated arrangement. We have similar cells that vibrate in response to sound waves and allow us to hear.

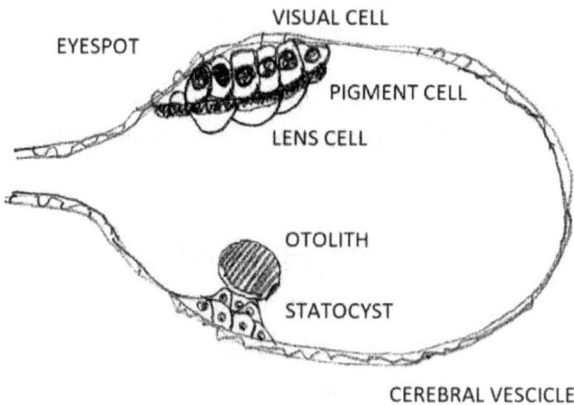

Figure 24 Tunicate Cerebral Vesicle: In the eyespot, clear lens cells transmit the light to the photoreceptor cells, and dark pigment cells direct the light path.

The action system consists of muscle cell controllers or motor neurons. They excite the first cells in three rows of muscle cells, which contract and set off the next set of muscle cells, and so on down the line. The tail sweeps back and forth, the notochord keeps the tail from telescoping, and the tunicate swims.

Between the sensory system and the action system, there is an interneuron processing system that transforms sensory signals into action signals.

Lancelet

At some point, the tunicate larva skipped metamorphosis to adulthood and became a free-swimming animal. This sort of thing is called pedomorphosis (child-form). It is not a deliberate step backwards; instead, a mutation makes metamorphosis impossible, or conditions do not allow it, and the organism is forced to live its life as a larva. In so doing, it grows up in a different ecological niche. If it has little competition, it can do well, replicate widely, and be a more successful organism.

Such an animal appeared about 540 million years ago. Its descendant, lancelet or amphioxus or branchiostoma lanciolatum, still swims about today. It is a free swimmer but does not swim much. When it does, it travels head down and close to the sea floor. It spends most of its time buried in the sand and filter feeding. It moves when there is no more food.

Its embryological development is like that of the tunicate.

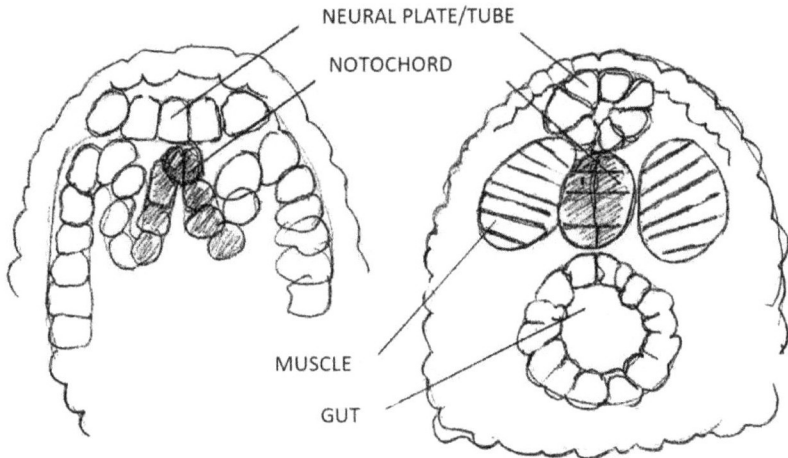

Figure 25 Lancelet Embryological Development: The neural plate becomes a neural tube. The dark notochord cells become the notochord proper. The folds to each side of the notochord cells become muscle cell columns. Once they and the notochord cells separate from the gut wall, they are no longer endoderm but mesoderm.

Its behavior patterns are more complicated than those of the tunicate, but not by much. It has two touch responses: If touched on the front end, it burrows further in. If touched on the back end, it swims forward. It has only these general escape behaviors. There are no more selective responses. It is a larval tunicate that moves when touched or out of food.

Lancelet Nervous System

Its nervous system is like that of the tunicate. The front end is much the same, with an eyespot and statocyst. The spinal cord region is better organized and more like that of the vertebrate. It is still a spinal cord animal with a tiny bit of brainstem function.

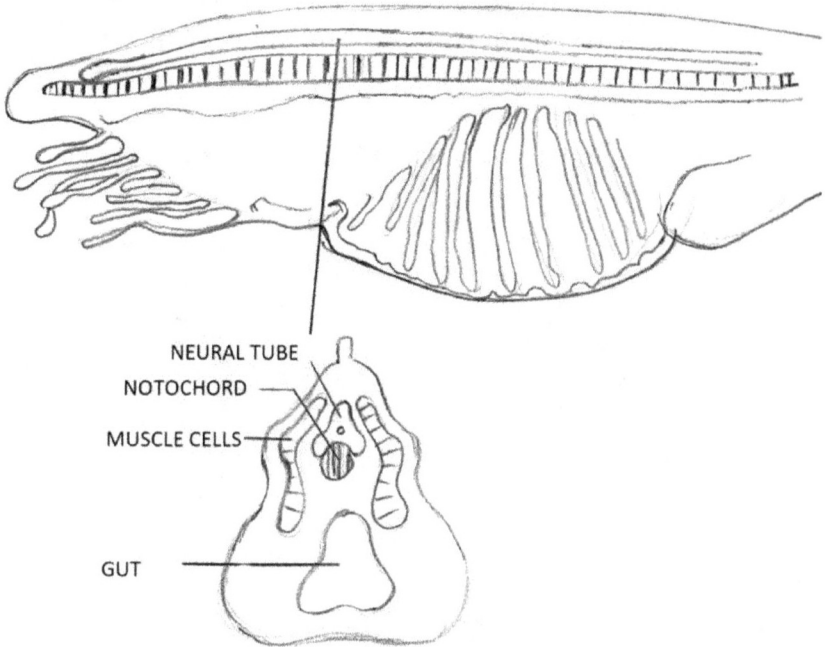

NEURAL TUBE
NOTOCHORD
MUSCLE CELLS
GUT

Figure 26 Adult lancelet in long and cross section. The tentacles at the mouth move fluid into the pharynx with gill slits.

The touch sensory neurons are towards the back and at the top of the spinal cord. The simplest touch receptor is a bare nerve ending. Pressure on the nerve ending destabilizes the membrane and triggers a stream of electrical discharges.

The motor neurons are at the bottom of the spinal cord with muscle cells sending in processes to connect.

This pattern of sensory input to the back and motor control to the front continues in all vertebrates.

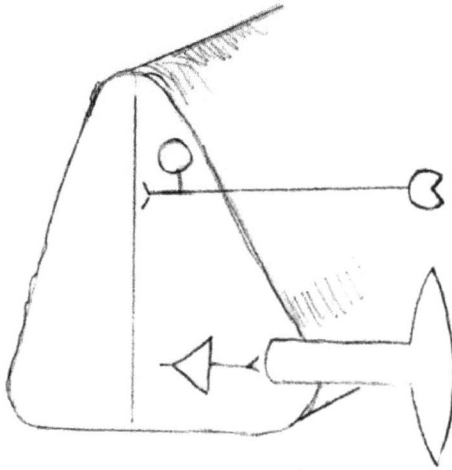

Figure 27 Lancelet Neural Tube: The sensory nerve cells are at the top of the cord and the motor neurons at the bottom. The sensory cells send receptors out. The muscle cells send processes in to meet motor cell axons.

As with the tunicate, there is not much forward of the spinal cord but there is the potential for there to be more. Vertebrates and their nervous systems are organized in segments with gene segments controlling the development of nervous system segments which control body segments. The lancelet has a single gene segment. Vertebrates have many. The gene segment was duplicated to make copies that mutated differently to make differing nervous system segments. The vertebrate pattern is latent in this primitive precursor.

The Block Diagram Model

The block diagram of the lancelet is like that of the tunicate. It has touch sensors, and a primitive motivation system that can register hunger and initiate movement to new ground.

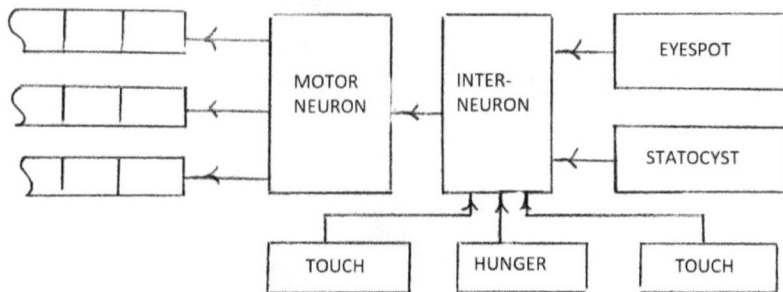

Figure 28 Lancelet Nervous System Block Diagram

The Neural Network Model

An engineer interested in modeling a nervous system as simple as that of a tunicate could choose on-off model neurons or summation neurons with ones and zeroes that work like on-off neurons.

The eye neuron connects to the motor neuron, which then connects to the first row of muscle cells. The statocyst cell serves only to inhibit the motor neuron by exciting a single inhibitory interneuron, which is all there is to the interneuron processing system. If the statocyst neuron was itself inhibitory, there would be no need for interneurons.

The model would act somewhat like a tunicate. A light signal would excite the receptor of the eye neuron, which would stimulate the motor neuron, which would stimulate the first row of muscle cells, which would contract and stimulate the second row, and so on. A wave of contraction would move through the tail, and it would move forward a little. If the light signal were still present at the next clock pulse, the swimming movement would continue.

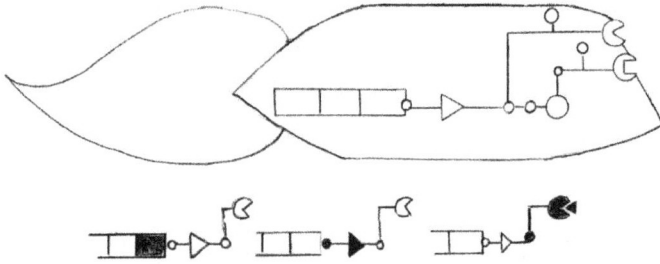

Figure 29 Model Tunicate: The light detector on top would activate a motor neuron which would activate the first muscle cell (the first of the three rectangles) which would activate the second one and so on. Below, a triangular light stimulus stimulates the light receptor and sensory cell which are shaded; then, the motor neuron; then, the first muscle cell. (The gravity detector is below the light detector and would excite the interneuron to inhibit the motor neuron.)

If the model sensory neuron, like a real one, needed a period to prepare itself to fire again, then the system would not need a clock pulse. It would fire at intervals determined by the refractory period whenever a stimulus was present.

If the gravity detecting neuron were set to fire only if the animal were pointing down and turn off the motor neuron, the model would drop straight to the bottom. It would never again make a flick of its tail unless a current happened to turn it eye-side up. Models seldom do as well as the real thing.

We could write the description or equation of the model's nervous system as

$$f = A. r$$

where the r's are the eye and statocyst receptor neurons, f is the motor neuron output, A is the set of connection functions.

This is a processing system: The dot operator operates on, or processes, inputs to produce outputs.

This is a stimulus-response system: Outputs are direct transformations of inputs.

This is an algebraic system: The connection function set (A) is the algebraic transform of sensory neuron inputs into motor neuron outputs.

We are using an engineer as a model maker because we are using an engineering approach. We are reverse engineering the chordate: taking it apart, creating a block diagram of its functions, and replicating them in the simplest possible way.

The Processing Perspective

The processing perspective is the view from thirty-thousand feet. It is looking at the nervous system as a processing system or abstract set of operations. It is looking at it as an engineering model, and from above all the noisy detail.

Detail is a problem. The sheer detailed complexity of a nervous system is a form of noise and a barrier to understanding. Noise is always present and a problem in engineering, particularly in sensory systems—smoke can get in your eyes.

From this elevated perspective, a nervous system is a sensory-motor input-output system: Sensation in, action out.

It is a sensory-motor processing system: Sensation in, processing within, action out.

It is a functionally split system: Sensory input above, and motor output below. Sensory processing above, motor processing below.

It is a bilaterally split system. One-sided touch activates a one-sided monosynaptic motor reflex. The crossovers of the upper parts of the human nervous system are absent.

It is a controlled bilateral system. Since the two sides must work together to swim, there must be a coordinating processor that sequences tail movement from side to side.

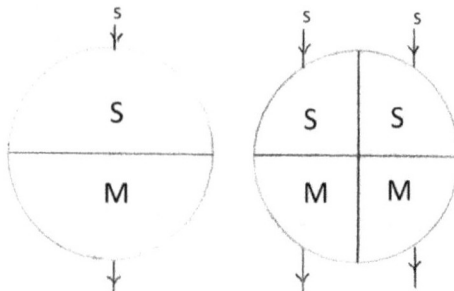

Figure 30 The sensory-motor input-output system and the bilaterally split system.

It is a multi-path system with different sensory inputs driving different motor outputs. Light drives not a monosynaptic but a polysynaptic swimming reflex.

It is a multi-mode system. Since the tunicate cycles through a sequence of behavioral modes, there must be a processor that switches on the light detector to swim up, then turns everything off to drift, and then turns on the gravity detector to swim down.

Processor and switch are terms for functions. They describe what a system does—a processing stage—but not necessarily a physical thing: A tail movement to one side could activate a reflex movement on the other side without the need for coordinating neurons. A stronger reflex for light could act like a switch without there having to be mode switching neurons.

So far, this description applies equally to pre-vertebrate and vertebrate nervous systems. The tunicate has a unique switch, a hormonal one, that flips on after adhesion and destroys most of the nervous system. This is not true of vertebrates, but some have hormonal switches that alter their nervous systems. We have puberty.

The Stimulus-response World of the Chordate: Von Uexkull's Umwelt World

These animals are reflex machines: They register a stimulus and respond. Their sensory cells are hard-wired to their motor cells. They are stimulus-response mechanisms.

They are spinal cord animals. There is no brain. There is no center to the nervous system. No area integrates and directs.

They have few sensory receptors, and this is a major limitation. The sensory receptor is the point of entry into the nervous system. If there is no receptor for a stimulus, the stimulus does not register and, as far as the nervous system is concerned, the entity generating it does not exist. Since the chordate's sensory receptors are few, it experiences little. Since an organism can respond only to what it can sense, it is largely unresponsive.

The term, Umwelt, was used by Jacob von Uexkull to describe the sensory world of such simple organisms. The German translates as environment (Um-around, Welt-world, pronounced um-velt), but the connotation is of an impoverished environment: Limited sensing abilities allow only limited registration of the world around: a dark minimal world dimly registered by the few sensory receptors.

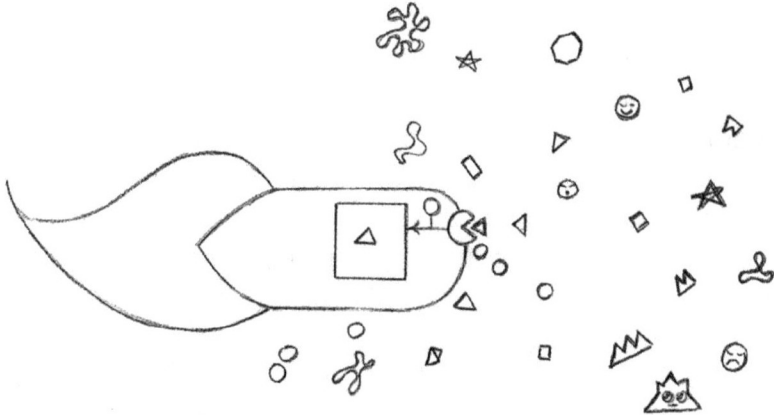

Figure 31 The Umwelt: The real world contains many interesting stimuli but the receptor registers, and the Umwelt contains, only one.

For the larval tunicate, the world consists of the upward pull of the sun and the downward pull of the earth. It is alone; its nervous system does not allow for the possibility of any other entity. For the adult tunicate, the universe has gone completely dark.

The lancelet or amphioxus can occasionally register a third state: a lightning bolt of touch which galvanizes it into frenzied activity followed by a return to quiescence. Most of the time, nothing happens.

The term Umwelt is misleading. It suggests a world view or world model, and, in these animals, there is none. Anything like a picture of the outside world will have to wait until much later.

> It's a long way from amphioxus. It's a long way to us.
> It's a long way from amphioxus to the meanest human cuss.
> It's goodbye to fins and gill slits. It's welcome lungs and hair.
> It's a long, long way from amphioxus but we all came from there.

CHAPTER 3

JAWLESS FISH AND SMELL BRAIN
AND DARWIN'S WORLD

When I get tired of looking at the sea squirts on the pilings of the dock, I can shift my gaze a few inches and look hundreds of millions of years into the evolutionary future where small fish swim in the eelgrass. There are a lot of fish down there. There are a lot of fish everywhere. The fish is an evolutionary success story; it is the most successful vertebrate in number.

The first animals with segmented backbones appeared around 400 million years ago. They were filter feeders that cruised along the sea-bottom straining for food particles. They varied in size from centimeters to meters.

They were the jawless fish or agnaths (a-absent, gnath-jaw). They were the first of the vertebrates. Their vertebrate backbones varied from cartilaginous notochord with boney thickenings around the spinal cord to true bony vertebrae. They had cartilaginous skulls and skeletons.

Their early fossil record is unclear since the early specimens were small and cartilaginous and unlikely to fossilize. The best recorded in the fossil record are the ostracoderms with armored heads, a later protective adaptation.

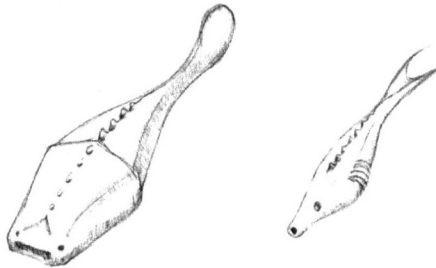

Figure 32 Agnaths: To the left is a three-foot-long ostracoderm (430-370 MYA) with an armored head and a single front nostril. To the right is a half-inch-long anapsida lasanius problematicus (there were problems in understanding its anatomy), without armor, and related to the later lampreys.

The animals in our developmental line have not been glamorous so far, and these mud-suckers were no improvement. They have evolved from mud sucking, but not to any great degree; their modern representatives are the lampreys and hagfish who suck blood from other fish.

There is, however, one amazing thing about these animals, best summarized by a classic Sidney Harris cartoon in which a scientist, who is having trouble getting from one set of equations to another, draws an arrow with the label, "And then a miracle occurs". Something like this happens with the agnath. It starts off as a spinal cord animal like the chordate, but then miraculously transforms itself into an animal with a brain.

We do not know how it does that, but we do know why. It does it so it can smell. We are going to look at how it smells and the world of smells it inhabits.

The Larval Agnath Lamprey

At two weeks the larva is fully developed. It is seven millimeters or a quarter inch long and looks very much like a lancelet. It stays in this state for five years, buried in the mud and filter feeding and growing to seven inches long. Like the lancelet, it moves only if it runs out of food and then swims head down close to the sea floor. This all seems very familiar.

Its nervous system does not do much more than the lancelet's. It has an eyespot on the head that is covered and non-functional, and another on the tail that responds to illumination by making the animal swim forward.

There is a change in the way motor neurons are organized: They send axons out to the muscle cells.

There is a change in the way sensory neurons are organized: Clusters of neurons develop outside the neural tube in crests on either side at the back and form the sensory neurons of the peripheral nervous system. The clusters are called neural crest placodes, and, in all vertebrates ever after, sensory neurons in the placodes send dendritic receptors out to register sensation and axons into the central nervous system.

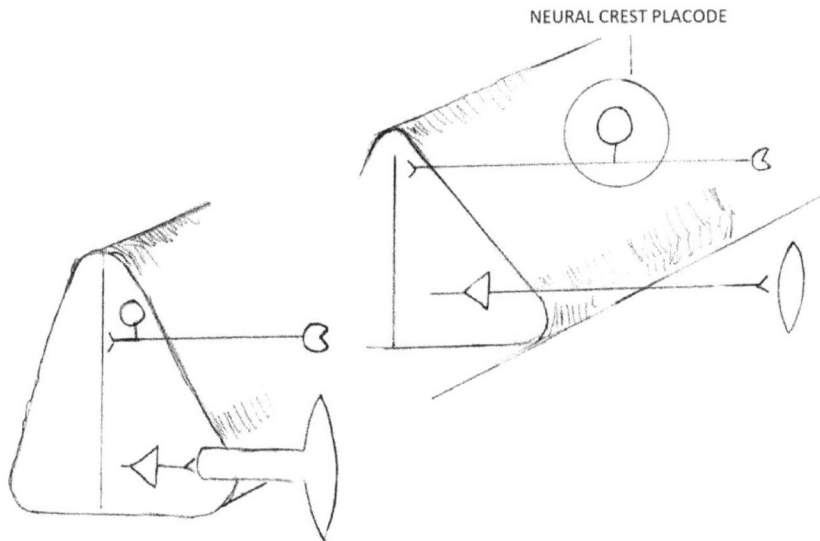

NEURAL CREST PLACODE

Figure 33 Spinal Cord Sensory and Motor Anatomy: To the lower left is the older chordate arrangement. To the upper right is that of the agnath and all later vertebrates, where the neural crest placodes contain sensory neuron cell bodies that send axons into the cord and dendrites with receptors out to the periphery, and where the cord motor neurons send axons out to excite muscle cells.

The placodes do more than direct the development of sensory nerves. The head placodes guide the development of the head and its smell distance sensors, and the development of the adult brain—the miraculous transformation.

Metamorphosis and the Adult Agnath

The larval animal is a simple filter feeder with a spinal cord and no brain. A chemical switch is thrown, and the miracle occurs. In the tunicate, metamorphosis turns it into a near plant with almost no nervous system. In the agnath, it turns it into an animal with the brain of vertebrates forever after.

The brain and its distance sensors allow the agnath to be a free-swimming predator. It can be thought of as a four-mode animal: In search mode, smell is its distance sense. A single nostril allows water to flow to the receptors of the smell nerve to detect molecules shed in the wake of its prey. When it closes in, it switches over to visual mode. When it touches, it switches over to touch and bite mode. It disengages and goes back into smell mode to do it again.

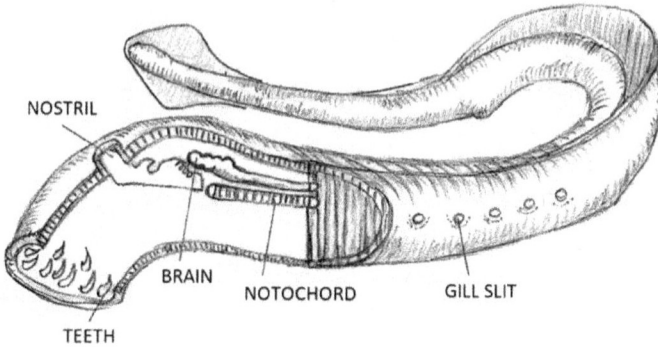

Figure 34 Agnath lamprey adult a foot long with cross-sectional schematic of the head and single nostril, brain and notochord. The mouth is circular and ringed with inward facing teeth. It is too large to do oxygen exchange through its skin like the tunicate and so uses its gills as a respiratory system.

The adult lamprey continues like this for two and a half years. Then another hormonal switch is thrown, and it changes again. It becomes a machine dedicated to making new agnaths. It stops eating, swims back to its home waters, reproduces, and dies.

The Adult Nervous System

The adult brain and brainstem have three functional areas:

The nerve placodes form the olfactory nerves (olfactere-to smell) and guide the development of the olfactory bulbs which receive the nerve signals. The bulbs send their signals to twin outpouchings of the brain, the cerebral hemispheres. The bulbs are larger than the hemispheres.

The visual placodes form the retinas of the eyes. The eye signals go to a processing area called the optic tectum (eye-roof) in what is called the midbrain.

Lower placodes form the taste and facial touch nerves that send their signals to the hindbrain, where biting and feeding are controlled.

The midbrain and hindbrain constitute the brainstem and largest part of the central nervous system.

The area between the hemispheres is a fundamental part of the brain but has another job. Its lower part is the pituitary gland that secretes hormones into the bloodstream to direct the hormonal glands of the body. It can be thought of as the brain of the hormonal system.

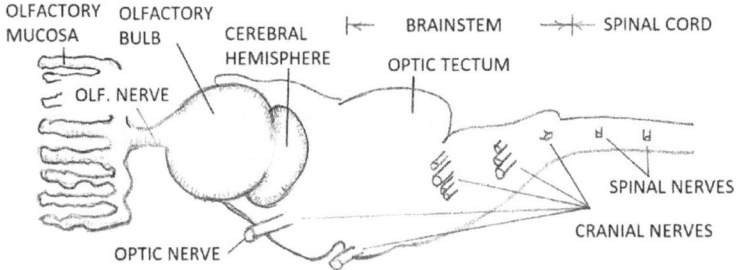

Figure 35 Agnath Lamprey Brain: The olfactory nerve in the mucosa (mucous membrane) sends signals to the olfactory bulb and cerebral hemisphere. The diencephalon is the large area between the two hemispheres with the pituitary gland forming the swelling below the optic nerve. The brainstem midbrain is next with the prominent optic tectum. The hindbrain with the other cranial nerves follows and then the spinal cord and spinal nerves.

The human brain follows the same basic plan. The hindbrain and midbrain are much the same. The pituitary hangs down below. The olfactory bulbs are similar. The cerebral hemispheres are greatly enlarged.

To clarify the nomenclature:

The central nervous system consists of the brain and brainstem of the encephalon (in the head), and the spinal cord (not).

The forebrain or prosencephalon (pros-forward), which consists of the central diencephalon (di-across) and the two cerebral hemi-spheres, is the brain proper.

The midbrain and hindbrain are older terms for the brainstem. The brainstem is transitional from spinal cord to brain: Its hindbrain functions like the spinal cord for the nerves of the face, but its midbrain has complex processing areas like the tectum.

The peripheral nervous system consists of the peripheral nerves connecting to the brainstem and spinal cord. There are twelve cranial nerves on each side of the brainstem in humans. The first is the olfactory and the twelfth controls the tongue. There are thirty-one nerve entry points on each side of the spinal cord in humans. The uppermost carry sensation from and signals to muscles at the back of the head. The lowermost is from the anus, not the toes. In a fish or an animal standing on four legs, the anus is at the end—unless it has a tail.

Smell or Chemical Distance Sensing

Smell is the distance sense. The olfactory neurons have processes that stick through the base of the skull into the nostril. They are studded with receptors.

The receptors are molecule detectors. They are protein molecules with complicated surface shapes that allow only mirror-image molecules to attach, fitting into the receptors like keys into locks. The neuron then sends an impulse train to the bulb to say that a smell target has been registered. The bulb sends to the tiny forebrain which sends to the brainstem to direct swimming to the target.

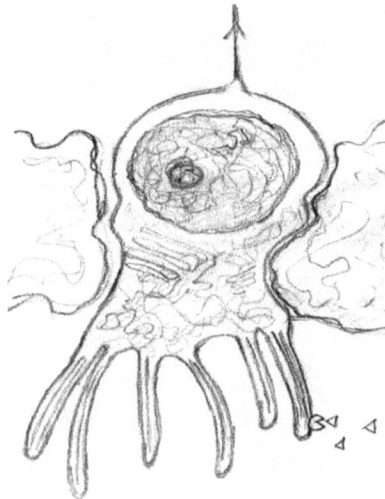

Figure 36 Schematic olfactory neuron with tentacle-like processes extending into the nasal cavity with a single olfactory receptor with triangular smell stimulus objects attaching. The processes are shaped by long tubular protein molecules.

The smell region directs until the target is close, and then turns control over to the eye and the midbrain, which then turns it over to the touch-and-taste hindbrain.

Taste is a chemical sensing system like smell. The taste buds on our tongues register a limited set of chemical signals: sweet, sour, salt, bitter, savory. Most of the taste of our food is smell.

The Brainstem Reticular Formation Control System

The smell-brain, and the taste and other brainstem sensory nerves, send their messages to the midbrain tectum. This in turn sends messages to a network of interneurons running through the brainstem that directs motor activity. It is called the reticular (net-like) formation and can be thought of as a switch that changes motor activity from smell-directed, to vision-directed, to touch-and-taste-directed.

The center of the central nervous system is the brainstem. It is the sensory and motor control center. The forebrain that is so important in humans is just part of the smell system: a cluster of interneurons that process chemical signals.

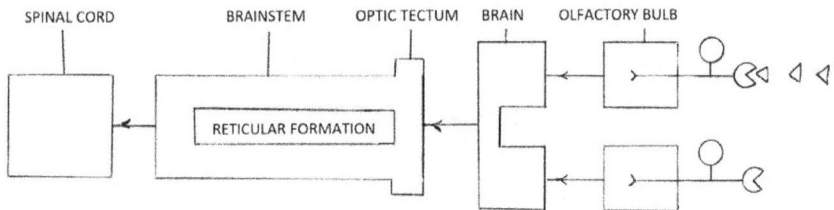

Figure 37 Block diagram of the adult agnath nervous system with an olfactory stimulus stream to the right activating an olfactory nerve neuron which activates the olfactory bulb which activates the largely olfactory cerebral hemisphere which activates the reticular formation. The two cerebral hemispheres are joined by the diencephalon. The eyes and brainstem nerves and connections are not shown.

The Jawless Fish Umwelt

What kind of a sensed world does the jawless fish inhabit? What is its Umwelt?

A smell world: Smell is the key distance sense for predatory fish. Vision is limited in the ocean. Touch is even more local.

Trailing behind each fish in the ocean is a chemical wake. The hunting fish swims through ribbon after ribbon of scented molecules: some fading and elusive; some pungent and arousing; some, seductive whiffs of dinner; some, red ribbons of danger.

Figure 38 The olfactory sensory world

A scent may never register. As has been said before, a stimulus from the outer world can only enter the nervous system through a sensory receptor. If there is no receptor for a stimulus, then as far as the organism is concerned it does not exist. The receptor set determines the sensed world.

The Processing Perspective and the Model Olfactory Predator

The smell chemoreceptor can be modeled as a binary-digit or on-off switch: If a molecule matches, it turns on and generates a one; if not, the receptor stays off, and its output is zero.

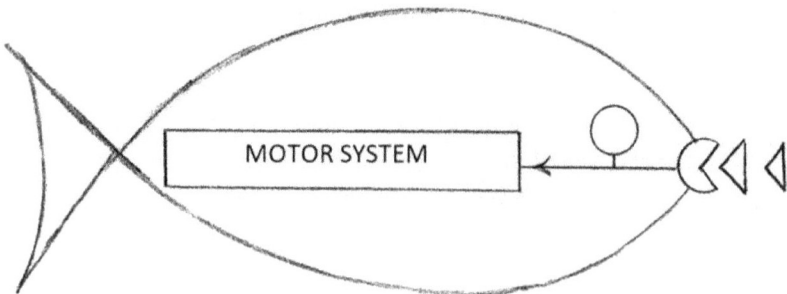

MOTOR SYSTEM

Figure 39 Model Olfactory Predator

The algebraic description of receptor processing would again be

f=A. r

but, if there were only one receptor, the connection functions would not be an array but a one, so that

f= r

and this would go to the motor system function box.

The model predator would need little sensory system; only one or more receptors. It would need no processing The receptor connections to the motor system would drive pursuit. It would need no memory. The functional memory would be the receptor set.

To pursue successfully, it would need more. It would need a motor system with a brain, brainstem reticular formation, and spinal cord motor system. It would need a switching control system. It would need an inner ear balancing system to keep it steady, a visual system, and an attaching and sucking system. From the processing perspective, we can ignore all of this.

Darwin's Model Ocean World

These model fish would inhabit a model ocean. The ocean would contain prey fish shedding one unique molecule, and predator fish with one receptor for one unique molecule. If the receptor detected a lot of fish, the predator would eat well.

The Darwinian organism has been called a replicating toolkit. The tools can be for action, like beaks or teeth; but for the model predators, they would be sensing tools, the smell receptors.

Evolution would take place through duplication of the receptor genes which would then mutate to make different genes and new variant receptors. If they were useful receptors, the predator would detect more fish to eat, and survive and breed. It would be selected, in Darwinian jargon, by its fishy environment to reproduce.

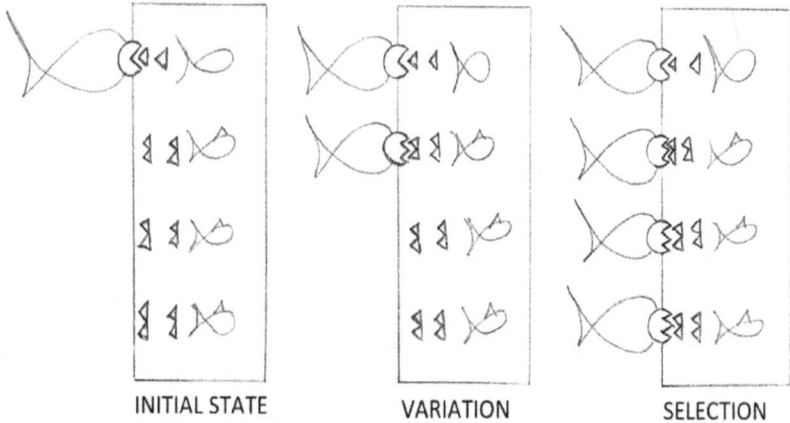

INITIAL STATE	VARIATION	SELECTION

Figure 40 The Darwinian Model Ocean: Genetic variation creates a predator fish with a new receptor. If the environment contains fish detected by that receptor, then the environment selects that predator fish for feeding success and reproduction. (The fish would still have the original receptor, but it has been left out for clarity.)

The Darwinian paradigm is adaptation to the environment through variation and selection by survival of the fittest. In this ocean, the individual predators are unable to adapt to their environment. They are fixed in structure, and their functional memories—their smell receptors—cannot change.

The system of predators is not fixed. Its system memory, the receptor set of all the predator fish, can change. Its system memory is subject to Darwin's paradigm.

We can use our usual equation in a somewhat larger fashion to model the entire system as

r= A. s

where (r) is a measure of the feeding receptors of all the fish in the ocean and (s) stands for all the prey fish stimulus molecules and (A) describes the interactions. The stimuli (s) are mapped onto the receptors (r), and if a fish's particular (r) is activated, then it can eat, survive, and reproduce. The (r) in the above equation could be thought of as a measure, not only of feeding, but reproductive success.

This ocean at any one time is a stimulus response mechanism.

The system response is determined by the stimuli and receptors. The future of the system is predicted by the equation.

The ocean over time is not a stimulus response system. Evolution introduces novelty into the system with each new receptor. It introduces double novelty since the Darwinian environment, the prey fish, can also evolve. The receptor set and the stimulus set can change.

The time course of the Darwinian Ocean is determined by evolutionary chance. It is a random process. It is a stochastic (stochos-chance) system. Its future cannot be predicted.

CHAPTER 4

JAWED FISH
AND FISHER'S PATTERN RECOGNITION WORLD

The jawless fish evolved from a muddy filter feeder to an active predator. Its role as a predator, however, was limited to nibbling. It could only slowly suck its prey dry. This confined it to a limited predator role.

The jawed fish appeared around 380 million years ago and was a major improvement. It had vertebral bodies made of calcified cartilage—bone—with slices of cartilaginous notochord forming disks between them. It had a boney skull and biting jaws that enabled it to become a successful ocean predator. The next 50 million years is the age of the jawed fish, some the size of a bus.

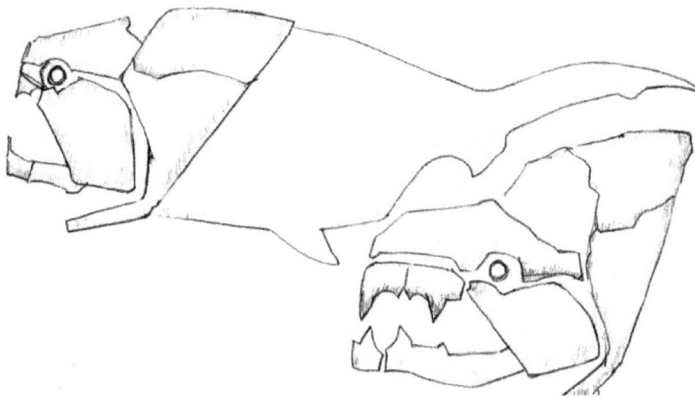

Figure 41 The predator placoderm, Dunklesteosis (382-352 MYA), was 20 feet long with a boney skull and neck cape.

The jawed fish made a basic improvement to the nervous system. It developed cells that made and filled up with a protein called myelin that acted as electrical insulation. Chains of these cells wrapped themselves around the axons and increased the speed of electrical transmission by ten times. A faster nervous system meant a faster predator.

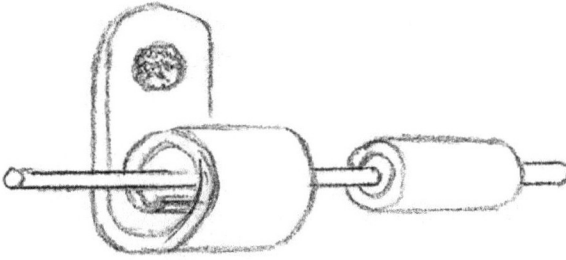

Figure 42 Myelin-containing cells folded around an axon. The leftmost cell with its nucleus at the top is shown in the process of wrapping itself around the axon. Next, is a bare axon gap. Next, the rightmost myelin-containing cell is shown wrapped around the axon. The electrical impulse hops down the axon from gap to gap.

It also improved the smell system. To operate well in its olfactory world, it needed smell receptors; anything it had to pursue or avoid, it had to be able to detect. The jawless fish had ten receptors; the jawed ones had one hundred. The mechanism of receptor increase was, as previously noted, duplication of receptor DNA with subsequent genetic mutation that resulted in variant gene copies and receptors.

The Devonian Age from 408 to 360 million years ago was not only the age of the fish but also the age of the olfactory animal.

We are going to discuss how its brain registered single smells and patterns of smells, and how it recognized those patterns.

The Olfactory Bulb Modules and Patterns

Smell chemical sensing is the simplest as well as the original distance sense and can serve to introduce sensory processing.

E.D. Adrian's 1950 theory was that a smell activated a spot or spots on the olfactory bulb dedicated to representing that odor. This is called a topographic (topos-place) system, the sensory information is displayed as a spatial pattern.

The receptors for each odor molecule send their signals to specific spots on the olfactory bulb where clusters of neurons form balls. There are 200 balls in the fish and 1800 in the mammal. They have pyramidal output neurons and smaller interneurons that surround them. The hundreds of output neurons and thousands of interneurons in a ball work together to act as a single unit or module.

The brain will later arrange itself in sensory modules and topographic displays in the touch, hearing, and seeing areas.

Sensory modules have three functional parts: input, processing, and output. In the bulb, the three functions are woven together. The pyramidal neuron is both input and output. Its dendrites receive the input signals, and its axon carries the output signal on to olfactory cortex. The interneurons do the processing work. They get their input signals from the pyramidal dendrites and send their outputs back through the same synapse. This synapse is unusual; most go only one way.

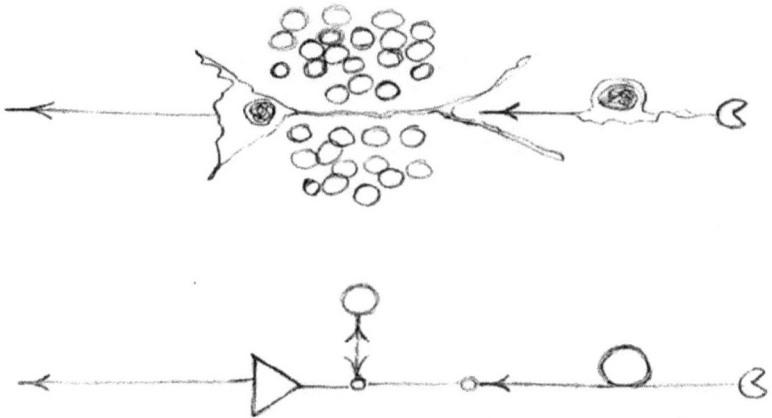

Figure 43 Olfactory module: At the top, an olfactory nerve neuron with a receptor connects to the dendritic tree of an olfactory bulb input-output neuron which is surrounded by a ball of interneurons to form a module. Below is a diagram of the module with a single interneuron receiving and sending through the same synapse.

The dendritic loop makes the module a feedback system. This term has taken on many meanings, some metaphorical, but to an electrical engineer it simply means that an output signal, a beep, is fed back into the system to re-excite it. Such systems can oscillate. If the feedback is positive, it amplifies the fed-back beep to produce a BEEP, and then again and again until the system shuts down or blows up. You have probably heard a microphone at a concert making a positive-feedback screech. If instead the feedback is negative, it reduces BEEP to beep, and then to something even smaller, and this continues until it dies away.

The olfactory module oscillates only when it is driven by the respiratory system. Each sniff induces a large electrical respiratory wave, and at the peak the modules burst into the forty cycle-per-second wave bursts of odor sensing. When the driving respiratory wave declines, the oscillations stop.

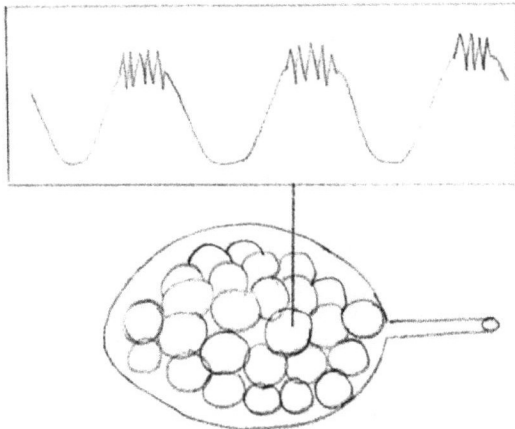

Figure 44 A schematic olfactory bulb with its ball modules. Only one module is electrically active with a large respiratory wave with a 40 cycle per second burst response riding on its peak.

The oscillations bind the activated neurons together to act as a chemical analyzer. A smell exciting only one receptor type would be a single hot spot; a complex smell would be a pattern of hot spots. The analyzer converts the chemical stimulus into a spatial pattern of electrical burst amplitudes.

Three-Layered Olfactory Cortex

The pattern goes to the olfactory cortex which has three layers like the bulb. How it works is less clear, but it represents smell signals and has smell memory. It sends its output signals to the reticular system of the brainstem to direct pursuit.

Its three-layer cortex has a synaptic layer, an interneuron layer, and an output neuron layer. This is the brain structure of fish and is still present in the older parts of our brains. It is called archicortex (archi—first, cortex—bark) or paleocortex (paleo—old). Mammalian brains are mostly six-layer neocortex (neo-new).

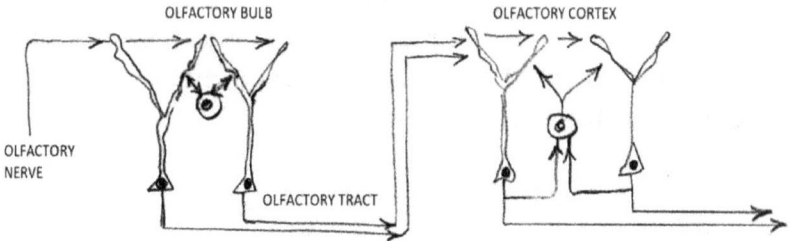

Figure 45 Olfactory Bulb and Three-layer Olfactory Cortex: The olfactory nerve connects to the bulb pyramidal cells, which form the output olfactory tract to the dendritic trees of the olfactory cortex. Bulb interneurons have two-way synapses that form feedback loops. Cortex interneurons have inputs from pyramidal cells through one-way synapses that form feedback loops.

Olfactory cortex also has oscillations that bind neurons in modules together to process simultaneously. Oscillations occur in the sensory areas of later organisms. They are not unique to sensory processing areas; there are oscillations in memory areas.

The Processing Perspective and Olfactory Pattern Coding

Each new receptor would expand the number of possible smells in the world of the fish but would also expand the number of possible combinations.

The model ocean of the last chapter was unrealistic. A real fish does not just drop one signature molecule for recognition purposes. It makes and releases many, and some are the same molecules that other fish release. The fish smell-signature is a compound signal. It could be considered a word with molecules for letters, or a number with each molecule represented by a digit. It is a pattern. A predator must be able to recognize the pattern.

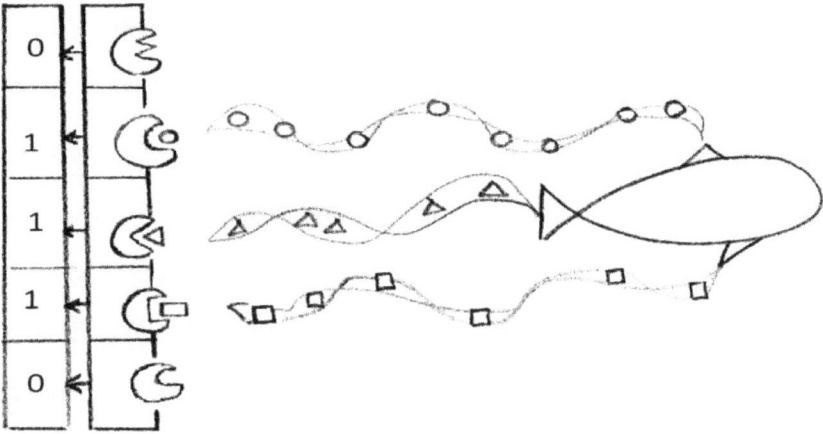

Figure 46 Olfactory Bulb Detector Model: The fish sheds three molecules and each is detected by a dedicated receptor and recorded as a one if present or a zero if not. The receptor response pattern is the binary number 01110.

This brings up the issue of representation of stimuli or how the nervous system codes its input signals. We do not really know, but it seems likely to be pattern coding.

A pattern is a multi-component representation. The simplest has two components: a black spot beside a white spot. This could represent one receptor occupied and another empty,

It could be coded in mathematical notation: [black, white] as [1,0].

It could be coded as a vector, a directed arrow in an x-y plot. The black-white or [1,0] pattern would be one unit along the x-axis with no y-component. A white-black or [0,1] pattern would be one unit along the y-axis. A black-black or [1,1] pattern would be one unit along both x- and y-axis. The last two-component vector is [0,0] or a vector with zero components and represented as a dot at the origin.

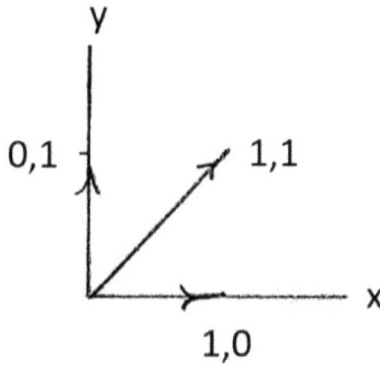

Figure 47 An x-y plot with the [1,0] and [0,1] and [1,1] vectors.

These are two-dimensional vectors. A three-dimensional object like a 2 by 4 piece of lumber that was 16 inches long would need a three-dimensional vector representation [2,4,16]. A floorboard would be wider and thinner [3,1,16]. Either could be represented as an arrow in a three-dimensional x-y-z plot.

We are confined to three-dimensional reality like the floorboard, but in mathematical reality you can have as many dimensions or vector components as you want. You can have four or four thousand.

You can code anything you want that way. Three symbols of any sort can be thought of as a three-dimensional object and coded as a vector. If a=1, b=2, c=3, and p=16, then [3,1,16] is "cap". English words can be coded as numerical vectors—or pieces of lumber.

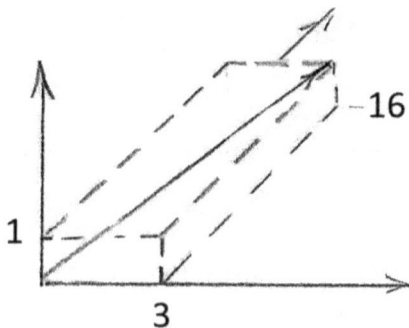

Figure 48 Three-dimensional floorboard with the third axis extending into the paper surface and the three-component vector [3,1,16] that could represent the floorboard or the word, "cap".

Our usual network equation in this notation would be called a vector transform

$f = A \cdot r$

where vector-r of firing rates for the input neurons is transformed into vector-f of firing rates for the output neurons. The array of connection functions (A) is called a matrix in vector algebra. It is a written description of the network transform and a mathematical convention for keeping the multipliers straight.

The dot operator here must do the tedious housekeeping of multiplying each row of the matrix by all the components of the input vector to get a single output component.

To be tedious briefly, we will transform the floorboard vector [3, 1, 16] with the matrix [1 0 0//0 1 0//0 0 1/4]: The first matrix entry [1 0 0] multiplies the first input component by one and the other two by zero to give the unchanged first component—tedious housekeeping indeed. The second produces the unchanged second input. The third multiplies the third input by one quarter. The transformed vector is [3,1,4], which could be the word "cad" or a floorboard fragment four inches long. The transform can code for a saw cut.

This is a lot of mathematical fiddling for a simple transform. We are not going to do any more of it. We will use network diagrams, and all we need to remember is that a vector transform is identical to neural network processing.

Non-vector coding is limited: one chemical signal for each fish, one letter for each section of the dictionary.

Vector pattern coding is versatile. It allows combinations. English with twenty-six letters can encode many more than twenty-six words. It can, in fact, code exactly twenty-six one letter words, but it can code twenty-six times twenty-six or 676 two letter words, and 17,576 three letter words.

A model fish with a single smell receptor can detect one smell, and its olfactory system has two states: on or off. It is a one-bit system with the bit either 1 or 0.

A two-receptor fish has four possibilities: no smells, smell 1, smell 2, or both smells. It is a two-bit, four-state system, written as a two-bit number (00, 01, 10, 11) or plotted as a two-dimensional vector.

For three-receptor vectors, the combinations increase to 2 times 2 times 2 or 8. For ten receptors, it is 2 to the tenth or 1,024. The increase is exponential.

In real fish it would not be exponential. Each new receptor would be similar to the last. It might recognize nothing new until several changes later. In testing real mammals with a thousand receptor genes, we can identify many fewer receptors. Despite this, more receptors and more receptor combinations markedly expanded the capability of the olfactory system.

Pattern Recognition Model

The olfactory bulb has been shown experimentally to recognize olfactory patterns.

We are going to make a model of the olfactory bulb that can recognize a pattern. In so doing, we are entering the esoteric world of mathematical models of the nervous system. This sounds forbidding, but we are going to make a very simple model.

There are three reasons for doing this: The first is to depict the nervous system in a new way, like a musical score as said earlier. The second is to suggest that its operations could use simple mathematics; that the brain is not an impossibly complicated and incomprehensible black box. The third is to demonstrate that mathematics can illuminate a brain system and reveal how it works, but the full demonstration will have to wait until the last chapter.

Our model ocean will have two chemically simple prey fish that shed only a one-molecule smell-signal. It will have one predator fish with two receptors (r1 and r2), so its receptor set response [r1, r2] to a fish 1 signal would be [1, 0] and to a fish 2 would be [0, 1].

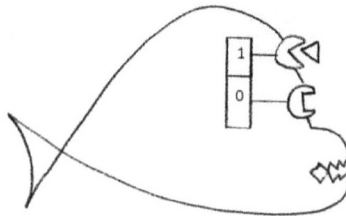

Figure 49 A two receptor jawed fish and its two-digit receptor response-pattern to a single molecule stimulus from prey fish 1.

The two receptors send their signals to two output neurons . There are four synapses linking inputs and outputs. This is an unrealistic model; we are ignoring the intrinsic processing cells.

The output f-neurons add their inputs after multiplication by their connection functions so that

f1= 8.r1 + 1.r2

f2= 4.r1 + 2.r2

Prey-fish-1 with input [1, 0] generates a large [8, 4] output, but fish-2 only a small [1, 2]. (I have a reason for these connection function choices that will become clear in the next chapter.)

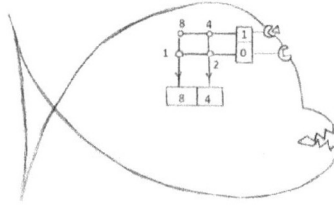

Figure 50 Two-receptor model with its pre-wired connection function network and its f-neuron response (8,4) to a (1,0) receptor input.

The olfactory bulb model turns the inputs into topographic patterns. It could be said to recognize the fish-1 pattern in that its response pattern is larger. The two patterns constitute the model fish's entire Umwelt.

This result is trivial. Of course, I can wire the connections to get an arbitrary numerical output. My point is that the nervous system can do so also; and, as we will see, can learn to do so.

Abstract Algebraic Network Model

We can translate the network model into vector algebra notation: We can write the f's and r's as vectors f= [f1, f2] and r= [r1, r2] and arrange the network connection values in a matrix (A), where

A= [8/4//1/2]

and the dot operator does the vector transform

f=A. r.

This describes what is happening, but the network diagram does so also and more clearly. The point is that the nomenclature collapses it into an abstract representation, a five-symbol sentence. It codes the network picture representation into an abstract algebraic representation, although this may seem an extravagant claim for five symbols.

Perception and Pattern Recognition

A network that processes sensory signals to accomplish pattern recognition is not just sensing but perceiving.

Perception is a word that has accumulated too many meanings for us to go into in detail. It is a step beyond the basic registration of stimuli. It is enhanced sensing through processing. It usually requires memory.

The sensory processing system now is a sensing system with multiple receptors followed by a perceptual processing system. The output of a perceptual processor is called a percept. The f-vector is a percept and is to be distinguished from the r-vector of the receptor set which is called sensory data or sometimes raw sensory data. The word, sensept, has been suggested but has never caught on.

We are going to confine our discussion of perceptual processing to pattern recognition of receptor responses. There is more to perception than this.

Pattern recognition is more of a mathematical than a neurological concept. The theory originated in the work of a statistician named Ronald Fisher in the 1930's.

Long before that, in the 1890's, William James had thought that a neuron only fired when it registered a particular object. This became known as the grandmother cell theory, that a specific neuron went off only in response to a specific grandmother. By the 1930's, it was argued that there were not enough neurons to register all the grandmothers and other objects in the world and that instead they had to be coded in some other way, perhaps as patterns.

The olfactory bulb model is a low-level, pattern-recognizing, perceptual processor. The one that is converting these marks on paper into representations in your brain is a high-level one. Perceptual pattern processors increase the capabilities of the receptor sets. They can be thought of as receptors for patterns.

The increase in ability comes at the cost of more ways to go wrong. In addition to a receptor failing to register a sensory stimulus, the sensory system can now suffer from a perceptual processor failing to recognize a pattern or getting it wrong.

Meaning

This simple model allows us to consider a deep subject: meaning. The larger [8,4] response means something—not much, but something: The model responds more to this than to that. One stimulus is more important than another.

We will come back to meaning again and again.

Model Olfactory Pattern Predator

We could hook the olfactory bulb model to the earlier model to make a pattern-recognizing predator model. A pre-wired setting in its perceptual processor would allow a pattern to be recognized, transformed into interneuron and motor system vectors, and it would swim.

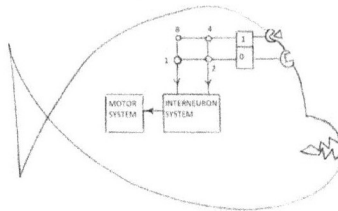

Figure 51 Pattern Recognizing Model Fish: The perceptual processor recognizes a pattern and activates the interneuron network that activates the motor system.

Unfortunately, it would swim away in all directions—or rather, wherever it was pointing. To go in the right direction would require steering and other systems. Again, there are always problems with models, but they are easier than making the real thing.

Fisher's Pattern-recognition World

The model fish swims randomly through a sensed world that is a featureless, dimensionless blur until a trail of chemical signals excites it.

Figure 52 The Olfactory World

Its Umwelt resembles a hospital corridor with colored lines painted on the floor: yellow to x-ray and red-blue to exit. If it hits a dinner line, it follows; if it hits a danger line, it turns away. It does not think—not even about what is for dinner. Its behavior is determined by its sensory inputs, as are its only decisions: pursue or flee.

It has a brain that is an early version of ours. That brain enables it to be a predator. The forebrain, in us a complex computing mechanism, is a simple odor processor. The center of the central nervous system is not in the brain but in the brainstem.

The model fish brain does not work like ours but does more than simple stimulus registration. It recognizes patterns of stimuli in the outside world—the task of the brain forever after. It uses the perceptual processing systems that will work so well in the evolutionary future.

It uses memory to do so. The pattern of connections that allow it to recognize a pattern in the outside world is a memory. Memory is a necessary condition for pattern recognition. We will deal with it in the next chapter.

The fish Umwelt is now a world of recognized patterns. It is an advance beyond the world of fixed receptor responses, but it is still dependent on them. It is still limited to an Umwelt determined by what its receptors can detect.

So are we.

Figure 53 Plato's Cave

We never experience reality directly; we experience the output of a receptor set. Like the prisoners in Plato's cave who can only look inward at the cave wall, we see the shadows cast by things in the outer world but not the real entities casting them.

CHAPTER 5

FISH AND LEARNING PATTERNS
AND HEBB'S CONNECTIONIST WORLD

We seem to be going on and on about fish—theoretical fish. We are going to stop. We are going to demonstrate that theoretical fish can learn patterns and be done with them.

Adrian's Model and Freeman's Olfactory Bulb Experiments

E.D. Adrian's theory of the smell system was that each smell triggered spots on the olfactory bulb that represented it and would always do so. The patterns, however, can change. The olfactory bulb can learn.

Walter Freeman in the 1980's measured electrical patterns in the rabbit olfactory bulb. When the rabbit sniffed and the bulb burst into 40 cycle-per-second oscillations, the amplitudes of the bursts were recorded with a grid of electrodes and displayed as a map of the responses.

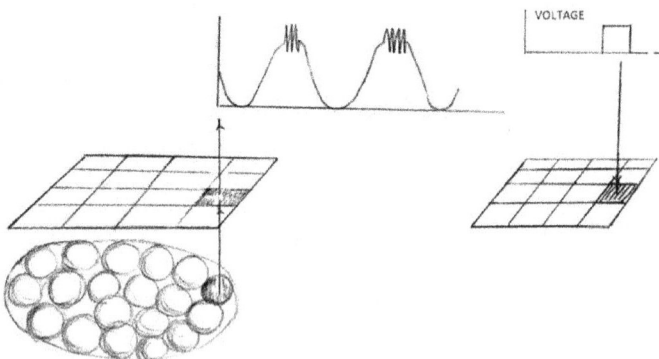

Figure 54 An olfactory bulb is shown lower left with only one module firing and generating an electrical response recorded by a grid electrode. Its 40 cycle per second burst response is converted to an average voltage amplitude of, for example, one volt and shown on the grid map as at lower right or written as the number (0000000000010000).

This was one of the first attempts to record large numbers of neurons in modules. The module patterns were expected to be Adrian's spatial images of odors, but the experiments did not work out that way.

Each rabbit turned out to have a baseline pattern like [1,0,0...]. It was present from the moment the equipment was turned on in the test-environment and remained the same for weeks when no odors were presented.

Each was then put through a Pavlovian conditioning experiment where an odor was paired with an electric shock. The bulb pattern changed to a new pattern like [0,1,0...]. This pattern was present from the moment the rabbit was returned to the test environment and through further sessions with the same odor.

Next the shock was paired with a second odor. The pattern changed to a new pattern [1,1,0...], again present from the moment the animal was put in the apparatus and through further sessions.

Finally, the first odor became the conditioned odor again and the pattern changed to a third and novel pattern [1,1,1...], but not to the previous first odor pattern.

If the pattern was there before the odor was presented, it was not an odor image. If it changed with conditioned odors, it was some sort of odor memory.

Freeman's interpretation was that the pattern was the odor that was next expected. The rabbit recognized the laboratory and set its bulb—before the experiment began—to register the conditioned odor. It was tuning its system to the environment. In the lab, this would prepare it for the next experiment. In the wild, for expected prey.

We are interested in fish. Jawed fish show similar tuning. So do humans; we are more likely to complete the letters, "hal... ", as halibut after reading a sea story. This is called sensory priming. It can be thought of as the effect of environmental context memory on stimulus memory.

Neural Network Model of Pattern Recognition

If the bulb can learn to recognize odor patterns, we need a model that can do that. It will be a neural network model with learning and memory. It will be as simple as possible and quite unrealistic.

We will use two receptors (r1 and r2) and two output cells (f1 and f2), and no internal processing cells, and set them so that each r-cell stimulates its f-cell a little more, so that

f1= 2.r1 + r2

f2= r1 + 2.r2

or

f= A. r

where f= [f1, f2] and r= [r1, r2] and A= [2/1//1/2] and the diagram of the two equations is

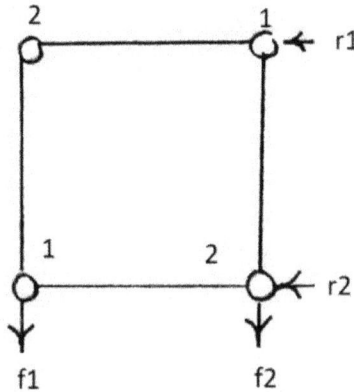

Figure 55 Pattern Recognition Simulator

A [1,0] input will produce a [2,1] output, and a [0,1] input, a [1,2] output. These are distorted or noisy versions of the inputs. This is not much of an accomplishment, but with learning interesting things happen.

The brain learns by changing synapses; the neural network by changing connection function values. The oldest and simplest rule for learning was suggested in 1949 by the psychologist, Donald Hebb, who thought that a synapse or connection function would change only if input and output were active at the same time.

Hebb also re-stated and improved on Cajal's 1894 hypothesis that memory resided in synapses. Time has been kind to Hebb and Cajal; there is now evidence to support both hypotheses.

Let us use mathematical notation to make Hebb's verbal rule explicit and model-able.

Let us decide that, as a measure of simultaneous activation of input and output, we will multiply them: r1 by f1 for the upper left connection value (a11), and r1 by f2 for the upper right (a12).

Let us decide that, to model learning, we will add them to the connection values: f1. r1 to its original upper left connection value (a11) and so on.

For a [1,0] input, only the upper (r1) connections get an input which yields the outputs 2 and 1, and only those connection functions change and learn by adding 2 and 1 to the original connection values to become 4 and 2.

If we have another [1,0] stimulus input, the output is [4,2]. and the connection functions change to 8 and 4.

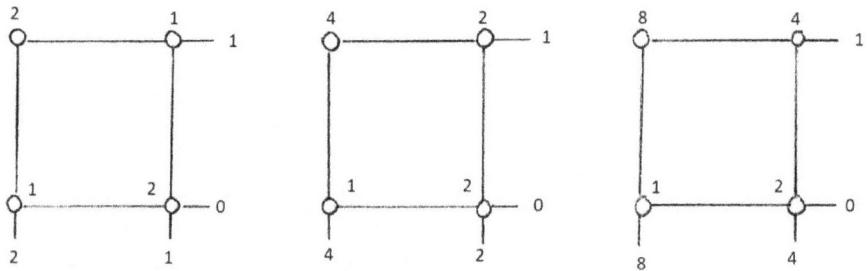

Figure 56 Bulb Pattern Recognition Simulation: A first [1,0] input to the left [2/1//1/2] network results in a [2,1] output. These values are added to the active connection functions on the top row to produce the changed [4/2//1/2] connection function pattern shown in the middle network. A second [1,0] input produces the [4,2] output shown, and results in the third [8/4//1/2] network on the right. That network's response to a [1,0] stimulus is the [8,4] output shown.

If we stop the learning process and check the system, a [0,1] input will excite a [1,2] output as before, but [1,0] will excite a larger [8,4] output.

Now we can say four things:

The network has learned the pattern recognition setting of the last chapter.

The network has responded more to a signal it has encountered before.

The network has adapted to its environment in real time and tuned itself to respond more to frequent stimuli; or, in the real world, to what is swimming around.

The network larger response has meaning: This signal has been encountered before.

Now we can say a fifth thing which is important enough to get its own heading.

The Network is a Model of the Outer World

The network is now a—very simple—model of the outer world. This is the point of the chapter and the book.

It has a memory and model of its stimulus encounters. The memory and the model are the connection pattern.

That is all it knows. Its stimulus encounter memory is its Umwelt, and all it knows of the outer world: There are a lot of [1,0]'s and not many [0,1]'s.

Olfactory Cortex and Top-Down Memory

The pattern in the olfactory bulb is sent on to the olfactory cortex. The pattern representation is no longer detectable there, and how it changes and what happens is less well understood.

The cortex is certainly a processing system and a memory system, and probably a better one than the bulb. Its neurons send both recurrent fibers back to re-excite themselves and recurrent loops back through interneurons. Recurrent loop or auto-association architecture, as we have noted before, could maintain a memory trace by sending a signal around and around the loop. It could also do so by changing synapse strengths.

The cortex also sends axon fibers back to the olfactory bulb. This is common in the nervous system. Sensory pathways going up are often paralleled by feedback pathways going back down. The top-down pathways are as important as the bottom-up input pathways.

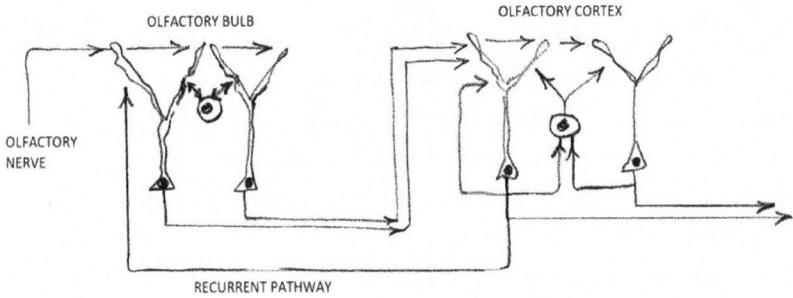

Figure 57 Olfactory Cortex and Bulb: The cortex pyramidal cells send back recurrent axon collaterals to their own dendritic trees. They also send axon collaterals to interneurons whose axons in turn connect to the trees. The pyramidal cells also send top-down axons back through a recurrent pathway to the olfactory bulb. Each cortex neuron would connect to both bulb neurons but only one recurrent line is shown for clarity.

The olfactory cortex is probably the place that stores memory patterns. A real fish would have to store not one but many preferred patterns according to where it was. The prey fish on the reef would be different from those in the eelgrass. Many stored patterns might interfere with the operation of the bulb. The cortex could store them instead and send down the settings for the bulb.

Other cortical areas could contribute. In later animals like Freeman's rabbits, the olfactory cortex is connected to frontal lobe areas that get inputs from other sensory systems. They could recognize an environment and signal the olfactory cortex to set the bulb pattern. This could explain the finding of bulb patterns before any odors were presented. The rabbit could have been setting its bulb when it saw the lab.

The Processing Perspective

The system now has the property neurophysiologists call plasticity (plastos-to form). It can form itself to suit its sensory environment. It can form new representations of that environment. It can form itself into a model of it.

Neural Networks and Connectionism

The network with learning is a neural network, and its memory is a connectionist memory—the memory is in the connection functions. It is, to be more exact, a connectionist memory with unsupervised learning. The network is not instructed or supervised by a teacher. It changes by itself in response to what it encounters. It passively records what it experiences.

Connectionism is a major concept in the world of neural modeling. Our connectionist results are modest, but our network is as simple as it can be—a finger exercise. An advanced one with multiple layers can recognize complex patterns and model complex nervous systems.

One other thing it can do is manufacture new representations.

Representation Combinations and Meaning

What good is recognition and memory to the model fish? What does it mean that the model can recognize a pattern?

It certainly does not mean, "Oh. That's a halibut." So far it just indicates that the stimulus has been encountered before.

We could add to the representation. We could associate a good taste signal with the smell signal. If we write it as a [1] in a mouth sensory receptor (and a bad as [0]), then we can send it to another bank of neurons where it combines with the smell input to form a combination or bimodal vector [8,4,1]. One sensory signal is now associated with another.

A smell signal by itself could trigger the combination vector. A partial representation could evoke the complete representation. Information from a past encounter, not present in the olfactory input signal, could affect the representation of that signal.

The combination vector expands the Umwelt. It expands the meaning. The combination signal would now mean good to eat, as well as smelled before. If it were part of a combination signal to the movement system, it could also mean that it should be chased.

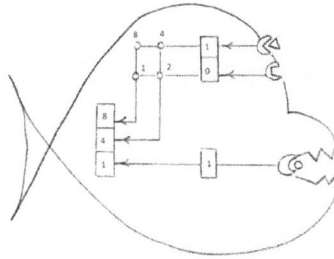

Figure 58 The association of smell and taste.

The meaning of "meaning" is expanding and will continue to do so as we move up the evolutionary ladder.

(Association has been used here in the psychologist's sense of learning to associate one stimulus with another. It can be modeled as the formation of an internal memory link between two stimuli. To replicate the Freeman experiments, we would have to model Pavlovian conditioned learning. We could do that by assuming that shocking an animal makes its learning changes larger. Perhaps the teaching establishment should not be told about this.

(Connectionist networks are sometimes called association networks because they associate an input with an output. We will stick with the term connectionist to avoid confusion.)

Distributed Representations

The vector representations we have been using are distributed representations. They are distributed over a set of neurons: Two in the example.

The memory of a connectionist network is similarly distributed over a set of connection functions: Two out of four store the memory in the example.

Distributed representations allow plasticity.

Distributed representations allow the manufacturing of new kinds of representations at a higher level, a kind of emergence. The percept combination is a symbolic representation of two different sensory representations.

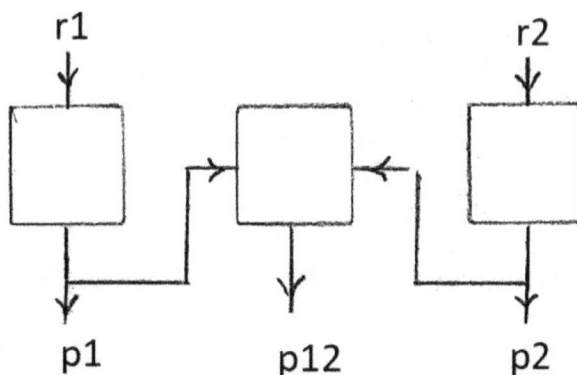

Fig 59 Two percepts p1 and p2 are transformed into a combined or bimodal p12 symbolic percept.

Distributed bimodal percepts improve perception. Two stages could allow two chances to recognize the input.

Distributed representations are used by real nervous systems. They are distributed not only over many neurons but also different areas of the brain like smell and taste.

The distribution property is important and has consequences.

Memory Damage Tolerance and Graceful Degradation

A real nervous system is not finely designed by a mathematician. It is a mishmash of sensing and processing systems cobbled together over eons. Its components are imperfect and noisy. It will have to work despite that. It will have to work even if it is damaged.

Catastrophic brain failure is unusual, even with catastrophic damage like a stroke; brains instead seem to suffer moderate impairment of multiple functions. The engineering term for this is graceful degradation; and, interestingly, a neural network pattern recognition system tolerates damage and degrades gracefully. The memory tolerates damage because it is distributed. Since many synapses change in response to the inputs, the memory is stored over much of the network, and a spot of damage will degrade many memory traces using those synapses but destroy none.

Input Fault Tolerance

Similarly, if the input signal to such a memory is incomplete or damaged, it still may activate enough synapses to retrieve the memory trace in its entirety. A partial input can result in a complete memory trace output as in the smell-taste example.

Tuevo Kohonen, a neural network investigator in the 1970's, took digitalized photos of people, probably his grad students, and trained his network memory with them. Then he fed in distorted or incomplete photos—in one case half the face was missing—and the network returned the complete photo.

The network demonstrated something like graceful degradation: When he doubled the number of photos in the network memory, it could still return the correct photos, but they were blurrier.

Noise Tolerance

Noise tolerance is another useful feature. If we add random noise to a signal and present it to the system ten times, then the cumulative signal will be ten times the signal plus ten times the noise. The signal term will add; but, since noise is random and a noise value of plus one at one time is as likely to be minus one the next time, the noise signals will tend to cancel one another out. If all you have are noisy signals—often the case in the real world—and you present many such, a connectionist memory will produce a clean version.

Generalization and Abstraction and Categorization

Variations in similar signals can be regarded as a form of noise. The memory system will not store the individual variant signals but an averaged or generalized version of them. Many particular halibut smells will become a generalized halibut smell. In later visual systems, many presentations of the color blue, will generalize to an abstract blue that encompasses robin's egg and navy.

The connectionist memory has a built-in tendency to generalize from the particular to the abstract and general. It can categorize. An impressive ability for a simple neural network.

New Problems and Dementia

We again have an increased repertoire of pathologies: In addition to receptor problems and perception problems, we have problems with learning as well as degradation or loss of memory.

The most common memory syndrome is dementia which is progressive memory and thinking decline. The commonest such disease is Alzheimer's dementia with degeneration and then loss of synapses and then neurons—the brain shrinks.

If the brain were like a neural network, it would suffer connection function loss with graceful degradation of its memories. Alzheimer's disease begins with, "the clouds coming down over the mountains", or difficulties with the highest of the intellectual and memory functions—perhaps problems with one's chess game. It progresses to problems finding one's way around the house or recognizing family members. It ends in apathetic confusion. The decline is gradual and somewhat graceful, at least at first.

The Hebb Model Ocean World

The Hebb world is a model ocean with prey fish shedding molecular odor patterns and predator fish perceiving them.

The model predator fish has a perceptual processor that can recognize patterns, as real brains do. The world, to paraphrase William James, presents a blooming, buzzing confusion of sensory signals. The brain's job is to find the meaningful patterns in the noise.

The processor passively trains itself to match the available patterns. The predator trains its internal memories with every encounter. It does not have to wait for the next generation to improve the perception system. It adapts to its environment in real time.

Its memory holds traces of everything ever perceived and the potential to hold everything that will be. Tennyson's Ulysses said, "I am a part of all that I have met". He might have added, "All that I have met is part of me".

Its memory is a model of its outer world, but an incomplete one, a limited model, an Umwelt.

Its memory and nervous system will be subject to more ways of going wrong. It will be subject to memory deterioration. It will be subject to learning problems: It may be incomplete or wrong. Taste may not be correctly associated with smell.

A critical problem will be learning the smells of bigger predators and surviving the learning experience. If it does not, it will not have the chance to link the predator signal to a flight program or do anything else. This suggests a need for anticipation and prediction, and we will get to this in later chapters.

PART II

VERTEBRATES ON LAND
AND SPATIAL MODELS OF THE WORLD

CHAPTER 6

AMPHIBIANS AND SENSORY TRANSFORMS AND YOUNG'S BRAIN LANGUAGE WORLD

Frogs are not particularly interesting unless you are in the Amazon where they are beautifully colored but also often poisonous. They do not seem to do much other than sit around and make noise. They are boring.

Frogs are like fish. There are a lot of them. They are everywhere. They are another evolutionary success story and dominate their amphibian niche.

Why do they just sit there? Why are they so successful? Why do they make all that noise?

It is easiest to explain why they just sit there. No matter how boring we would find sitting like a bump on a log or a lily pad; it is, for a frog, a successful strategy. They are sit-and-wait predators and good, if unexciting, ones. Flicking out the tongue to catch an insect does not seem exciting either, but the movement could not be simulated by a roomful of engineers with computers.

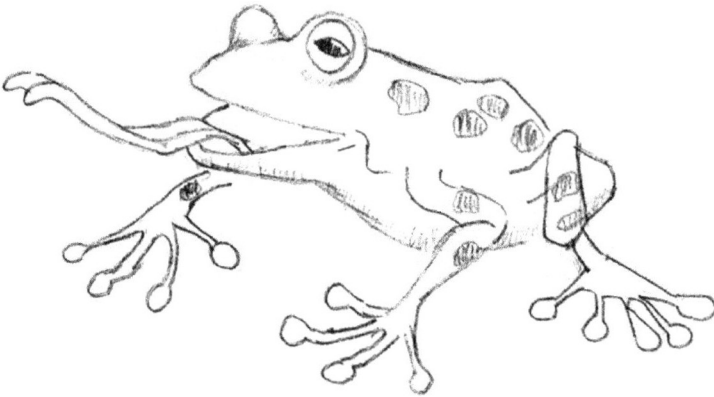

Figure 60 The frog's tongue can be extended to one third of its body length and is forked and pliable and coated with sticky saliva that can hold onto objects heavier than the frog itself.

The frog must be able to see to catch its insect prey. We are going to discuss the frog visual system, and the representation system it uses for vision, and also for smell and memory.

Coelocanths and Amphibians

About 350 million years ago, the fish evolved into a land animal. The fish that did this was not a speedy, successful sea predator; it was a heavy-bodied, unsuccessful cousin. It descended from the coelacanth which was thought to have been extinct until a specimen turned up in a fishing net in 1938. The coelacanth is a predator, but one that tends to sit and wait for its prey rather than chase it, and one with the atavistic habit of hanging head-down near the ocean floor.

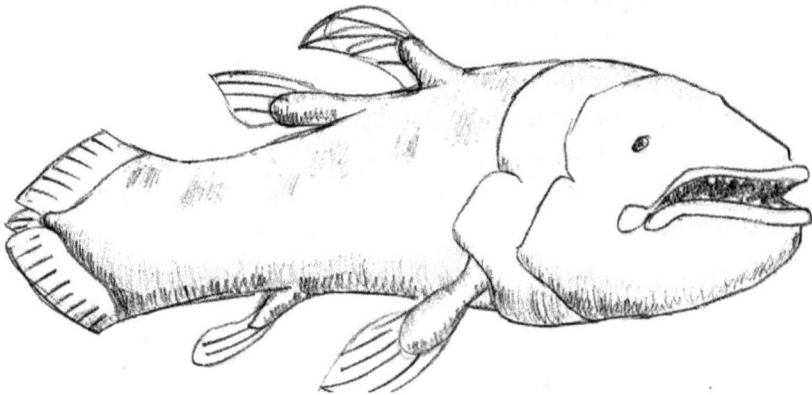

Figure 61 The lobe-finned coelacanth, latimeria, 3 feet long, and still swimming about in the depths of the Indian ocean.

The coelacanth is a living fossil. It never changed, but one of its descendants did. It occupied a marginal niche in shallow coastal water where it tended to get isolated in tidal pools. It used thick lower fins to slog through the mud to the next tidal pool and used internal bags of air called swim bladders to get oxygen into its bloodstream when its gills were not moving water.

It began to spend more time on the tidal flats. The reason was food. It followed the insects that had moved onto the land millions of years before and flourished in a world without predators. In the water, it scrabbled for scraps; out of it, the living was easy. It followed the path of easy energy—of biological wealth—up onto the land. Its fins evolved into legs and its air-bladders into lungs.

There was not much on the land other than insects. Plants were small and could survive only on the shore with their roots still in the water. The rest was bare rock.

It did well there. The slow cousin finally had a decent job. It multiplied, expanded its biological range, and changed to adapt to land, but not that much. It continued to breed in the water, laying its eggs the old way. It spent its early larval years as a filter-feeding tadpole, living the old way. Only after metamorphosis did it move onto land. Once there, it still acted like a fish—a fish out of water. In the water, it swam with speed and agility. On land, it wiggled from side to side like a fish on stilts. It was ungainly and slow.

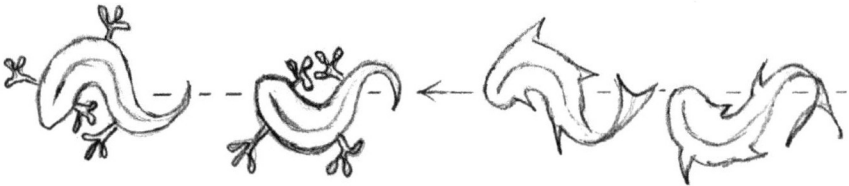

Figure 62 Swimming to walking simplified.

In the water, it tracked chemical trails with a wet chemical sensing system. On land, there were only intermittent whiffs of air-borne scent, and its system did not track as well.

It really was a fish out of water and about as efficient. Luckily, that did not matter. Food was abundant. It did not have to track it; all it had to do was sit until an insect came by. It could get along as a sit-and-wait predator; and it did not have to worry about other predators—there were none, at least for a while.

One of the earliest amphibians, icthyostega, looked like a walking fish with a fishy tail as well as legs.

Figure 63 Icthyostega: A 5-foot-long transitional amphibian (374-359 MYA).

Its descendant, the frog, does not look like a walking fish but has not changed its behavior all that much: It lays eggs in water and its tadpole lives as a filter-feeding swimmer for a season. It metamorphoses, finds a lily pad, and snaps its tongue out at passing flies. It makes all that noise. Why?

It is a signature signal, as Emily Dickinson knew and said in a poem about celebrities:

> How dreary to be somebody!
> How public like a frog
> To tell one's name the lifelong June
> To an admiring bog!

And it is all about sex. It is a mating call.

The amphibians are a transitional group of animals. Some breathe through the skin like fish and others through the lungs like reptiles. Some fertilize eggs in the water like fish and some internally like reptiles. Some are retiring and others self-advertising.

Olfactory Memory and What the Frog's Eye Tells the Frog's Brain

The amphibian had the fish smell sense, but it had a problem the fish did not: The fish tracked continuous, chemical trails in water. The amphibian got intermittent whiffs of air. It had to remember what it had smelled before to make sense of the next whiff. It had to rely on olfactory memory.

As amphibians evolved, they would select for better smell memory. However, since they were still stuck with the essential limitations of air scenting, they would have to rely on other sensory systems. Since their insect food was rapidly moving and often flying, they would select for vision with signals moving at the speed of light rather than wind.

The frog brain is the same size as the fish brain but has changed to develop useful land vision. This is not vision of our sort though. We send visual signals to the cortex of the brain. The frog sends them to the upper brainstem like the fish does, but that optic area is bigger and more specialized.

In the 1950's, a group, which included the same Warren McCullough of the first formal neuron model, stuck electrodes into the frog's optic region to find out what the neurons responded to. Some responded to small moving objects: they were bug detectors. Others responded to sudden changes in overall illumination: they were overhead predator detectors.

In a classic paper called, "What the Frog's Eye Tells the Frog's Brain," they argued that this visual system was not concerned with pictures of the world outside. It was a stimulus detection system tuned to the significant issues of the frog's life: eating and getting eaten.

The Frog's Brain

The frog's brain is still a minor part of its nervous system. The important part is its upper brainstem.

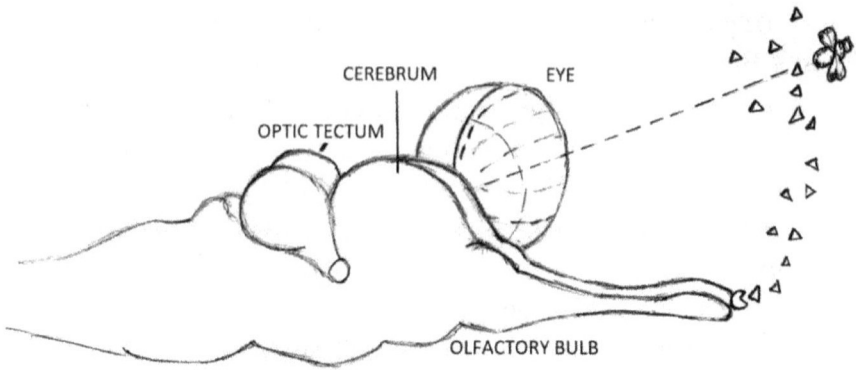

Figure 64 Schematic of a one-eyed frog brain with an optic tectum or visual area in the upper brainstem that is almost as big as the brain. The left eye is shown as a retinal posterior hemisphere with the lens and front of the eye missing. A flying bug is a double stimulus, dropping smell molecules onto the olfactory nerve receptors and reflecting light onto the retina.

The receiving area for visual signals is the optic tectum in the roof of the midbrain near the upper end of the reticular formation.

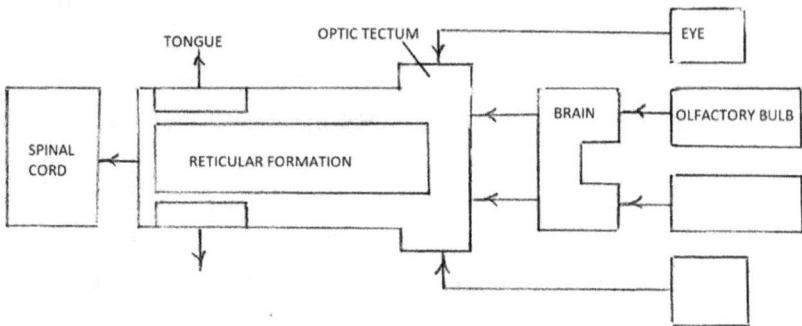

Figure 65 Frog Brain Block Diagram: The eyes connect to the optic tectum of the brainstem. The olfactory bulbs connect to the olfactory cortical areas of the brain which connect to the brainstem and the reticular formation.

The tectum is the sensory center of the nervous system. The optic part receives visual information. The lower part receives touch and hearing information which is kept in register with the visual information.

The signals are sent to the movement centers. An area below the tectum directs eye and head movements. It is a continuation of the reticular formation which directs tongue movements further down the brainstem.

The Visual to Tongue Control System Model

The optic tectum detects small moving bugs and sends signals to the tongue control area to fire the missile-like tongue at them. The tectal location must be translated into tongue muscle direction control. The sensory signal must be converted to a motor signal.

There is an old joke about an engineer writing a biology paper who begins with, "Assume a spherical chicken." Let us follow his lead and assume a squashed-flat, one-eyed frog that only goes after bugs in its 2-dimensional plane.

The engineer would break the problem down into solvable parts. The first would be the standard problem of different reference frames. The retina would specify an angle to the target like forty-five degrees. The tongue control system would not and would need rewritten coordinates. The second problem would be tongue muscle control.

Assume the tongue uses two-dimensional x-y coordinates: Ten across and ten up would correspond to forty-five degrees.

Assume a connectionist network to transform the angle location vector into an x-y coordinate location vector.

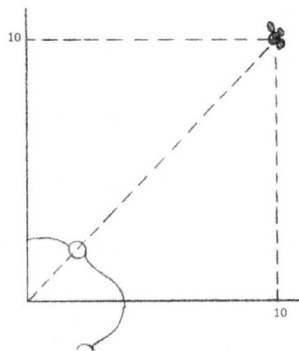

Figure 66 Model Reference Frames: The head of the squashed-flat one-eyed frog has a bug at 45 degrees in its visual coordinate system and at x=10 and y=10 in its x-y tongue coordinate system.

Assume another transformation into a muscle contraction vector. This would be the hard part of the problem, writing down a vector that activated multiple muscles, all contracting at once, to launch a flexible whip to a spot in space. We will leave this to the roomful of engineers with computers.

The Frog Brain Block Model

We could convert our fish model to an amphibian model. Three motor control pathways would respond to their specific sensory inputs: The olfactory bulbs would detect olfactory targets and direct head and eye movement to align with the stimulus. The bug detector system would detect small moving objects and direct tongue motor control. The overhead illumination-change system would detect threats and direct leg jump control.

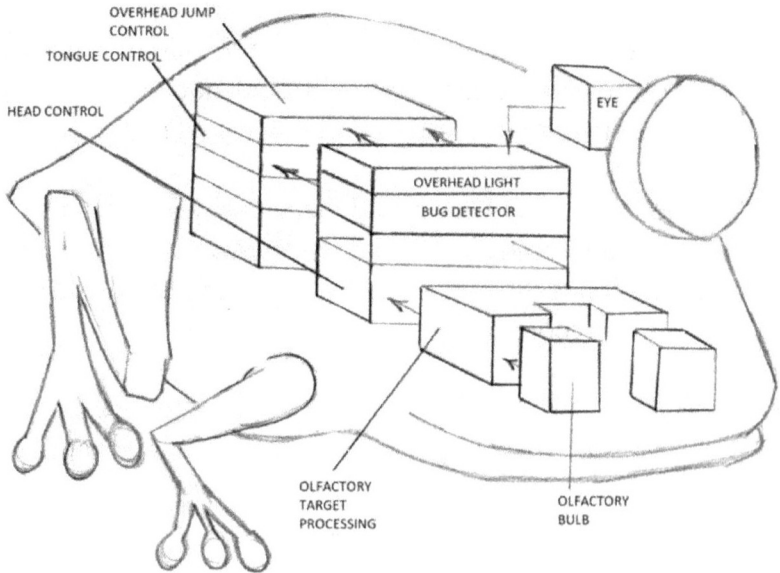

Figure 67 Unsquashed One-eyed Frog Model: The brainstem is the two large boxes. The olfactory brain behind the olfactory bulbs connects to the brainstem head and eye direction system. The single eye goes to the bug-detector and the overhead-light-change detector in the tectal area.

The Brain as a Coding Problem

We have described a sensory signal as a word or number. We have described combining a smell signal with a taste signal to form a compound word or number. We have described a transformation of a compound word and number like forty-five-degree-angle into another like ten-over-and-ten-up.

The zoologist, J.Z. Young, thought that brain function should be studied as a language coding problem. That sensory stimuli are coded into receptor representations like words and written into the brain in a kind of language. That the words are transformed into different words according to the rules of that language. That the job of the brain researcher is to sort out the language.

Unfortunately, in the brain, we cannot do that.

Fortunately, in the brain model, we can. The algebraic model is

$$f = A. r$$

with brain words that are numerical vectors and processor operations that are vector transforms.

The model can code the sensory processing operations of the model brain as a sequence of transforms resulting in a brain activity vector (b)

$$b = C. f$$

which is transformed into brain output activity (x)

$$x = B. b$$

which acts on the brainstem movement system.

This may seem overly simple, but brain language must be something like this. Our millions of sensory receptor values could be written as a large numerical vector and translated into a brain vector. The activity pattern of the entire human brain could be written as a gigantic vector with 100,000,000,000 neuron values or

$$b = b(n1, n2,..., n100,000,000,000).$$

This brain activity vector is a mathematical function. It is a function of these neuronal activities, of their synaptic activities; and, since they all change with time, a function of time.

The function is a fundamental mathematical concept:
It is an abstract representation of a system.

It is an abstract representation of an action. It specifies how a mathematical object is processed or transformed: How inputs are mapped onto outputs. How vector-r becomes vector-f.

It is an accounting tool that keeps track of the controlling variables: That vector-f is a function of and so determined by vector-r.

Keeping track of controlling variables is a major concern for mathematicians and for neural modelers. It is a major concern for all of us. We are all functions of our controlling variables.

The Processing Perspective and Learning

We can put the learning model of the last chapter into this notation. We can code the controlling variables. We can map them through learning into memory.

For a single synapse (a) where

$$f = a.\, r$$

after an input (r') and output (f'), the simplest Hebb connection change (Δ a') would be

$$\Delta a' = f'.\, r'$$

so that

$$a' = a + \Delta a'$$

and similarly for a vector with a connection matrix (A)

$$f = A.\, r$$

and

$$\Delta A' = f'.\, r'$$

and

$$A' = A + \Delta A'$$

where the $\Delta A'$ is of an array of Δ a's, and the vector dot operator now does the multiplications and puts them in the right positions. (As Humpty Dumpty might have said, "When I use a dot operator, it means just what I choose it to mean.")

After another input r''

$$\Delta A'' = f''. r''$$

and

$$A'' = A + \Delta A' + \Delta A''$$

and this can go on until

$$A'''' = A + \Delta A' + \Delta A'' + \Delta A''' + ...\text{and so on,}$$

and now we can say:
Processing is A (and then A' and so on).
Learning is ΔA' (and then ΔA'' and so on).
Memory is A + ΔA' + ΔA'' + ΔA''' + ... and so on.

The Amphibian Multi-Representation and Young's Brain Language World

The early amphibian nervous system was not a major improvement on that of the fish, but the demands of living on land would select for more sensory processing and more memory.

The Umwelt was split into two. The olfactory system in the brain would require memory and processing time. The visual system in the brainstem would be a stimulus-response system. The two would compete for control in the brainstem.

In J. Z. Young's brain-language world, representations are like numbers or words, and communication between areas requires transformations of numbers or words.

The amphibian model can transduce, code, remember, and transform sensory signal representations:

The model of coding is vector representation.

The model of memory is connection function change.

The model of sensory processing is vector transformation.

The model of motor control is vector transformation.

The model of the entire brain is a sequence of vector transforms.

The model is a few lines of linear algebra.

CHAPTER 7

EARLY REPTILES AND BIOLOGICAL CONTENT ADDRESSABLE MEMORY AND KANT'S MAP WORLD

I lived for a while in east Texas near the Louisiana border in a house buried in the trees of a deer park. The second-story balcony was the best room in the house. In the spring, pollen fell like rain, and I would sit there in a totally green world, waiting for deer and watching green lizards plowing through the pollen and brown lizards clinging to the wood. Eventually, I realized that they changed color and were the same lizards.

Within minutes of my sitting down on the balcony, a lizard would appear on the railing, stare at me, puff out his throat pouch, and do a push-up. This happened over-and-over again, always the same display. Eventually, I saw one lizard make this display to another, and then climb onto its back and bounce up and down for a while. Then I understood that the lizards were communicating and what they were communicating, but it did seem a little odd that they were bothering with me.

Figure 68 Anole lizard display.

This display sequence is, oddly enough, important. It is a progression from the stereotyped here-I-am noises of amphibians toward more complex communication.

We are going to discuss communication and the appetites and aggressions that drive it. We are going to discuss the hippocampus and amygdala where these things go on. We are going to model the memory system.

The "Early Reptiles"

Three hundred million years ago, improvement of the egg freed the amphibians from the water and created the first true land animals. In the textbooks, they are called amniotes after a component of the egg that lets them reproduce on land. They evolved into reptiles and mammals.

They soon separated into three groups, named according to the number of holes in their skulls in addition to the nose and eye sockets: Those with none were the earliest and are called anapsids. The turtles were once thought to be surviving members of this group, but this is now disputed, and the group thought to be extinct. The diapsids had two and gave rise to reptiles like dinosaurs and lizards. The synapsids had one and gave rise to the mammals.

We are interested in the early basal amniotes that evolved into synapsids and then mammals. We could use that awkward name, but since we are going to use present day reptiles as the only available examples of what they were like; and, since the basal amniotes were early and reptilian, we will just call them early reptiles.

Since they differed from the amphibians primarily in the nature of their egg, it has been difficult to tell them apart in the fossil record and be sure when the two species separated but it is thought to have happened around 312 million years ago.

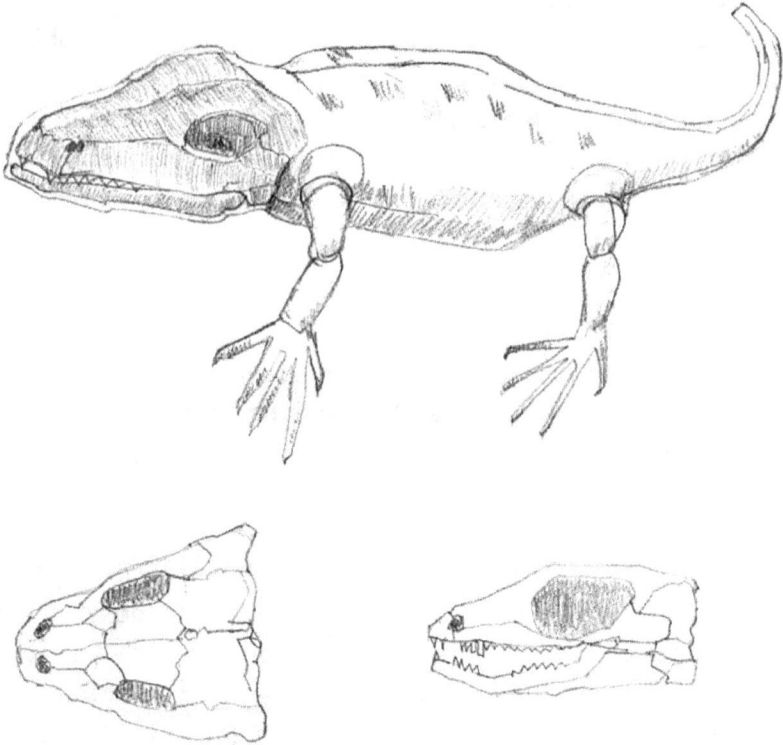

Figure 69 The first unquestionable early reptile, Hylonomus (312 MYA), 8 inches long, with a triangular head and small teeth for eating insects.

The land was now somewhat more inhabitable. Plants had lost their need to have their roots in water and fern forest had spread over swampy land.

The plants evolved. The first pollen-dispersing trees, the conifers, developed around 250 million years ago, and pine forests began to cover the land. The world began to look more like the one we see today, but there were no flowers and no leafy trees.

The first land invaders had the run of the place for a long time. They evolved and diversified into plant-eating and flesh-eating animals. Some became search predators rather than sit-and-wait predators.

The search predator niche rewarded animals with better sensing, memory, and motor control. Their behavior and their brains improved. So did their teeth—an early form of arms development.

The early reptile was a hunter, but a plodding one. It was a creature of rigid habits: The same daily routine; the same stereotyped displays; the same hunting program. It did not innovate. You do not have to worry about present-day reptiles figuring out how to get into the bird feeder, but you rack your brain to keep out the squirrels.

It had to deal with two problems the amphibian did not: It moved about on land and had to use its brain to find its way. It had to fertilize its eggs internally, which required that males and females co-operate, which required more complex communication.

They were non-vocal animals, and communication was through body postures. For the anole lizard of the southern United States, the primary display is called the signature display, and consists of head-nods and push-ups with a brief flaring of the throat pouch. This seems to mean something like, "Here I am."

The courtship display of a male to a female is an elaborated version of the signature display that induces the female to go into a crouched posture that allows the male to crawl onto her back, bite her neck, and copulate. This does not sound like much in the way of male-female communication, but it was a beginning, and it has not entirely gone away: Think of female hip tilts. Think of male hands-on-hips stances. Think of Elvis.

Lizards are territorial, and they signal that their territory has been intruded upon with a threat display. This is also based on the signature display but with more aggressive flaring of the throat pouch and the erection of a ridge on the back to look bigger. This display can be countered by a return threat display from an aggressive animal, or by a submissive display with a down-going head by a non-aggressive one.

In retrospect, this was the display I was getting—they were Texan lizards after all.

The Early Reptilian Nervous System

The early reptiles do not exist anymore. As far as brain function and behavior go, everything I am going to say is speculation. I am taking reptilian characteristics and dating them back to a hypothetical early amniote in the process of becoming a synapsid.

The early reptilian brain was, and the reptilian brain is, no larger than that of the amphibian. It did not need a bigger brain, but it did need a different one.

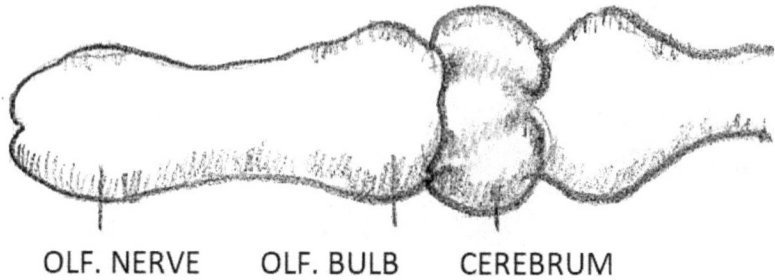

OLF. NERVE   OLF. BULB   CEREBRUM

Figure 70 Tyrannosaurus rex 6-inch-long endocast, with olfactory bulb, cerebrum, and optic tectum bulge behind, all about the same size.

The olfactory bulb receptor gene set expanded, but not to the full mammalian set of 1000 genes. The early reptile probably had more receptors and better smell memory and would have experienced selection pressure for better olfactory function.

Vision and hearing improved, but not that much. Both senses still reported largely to the brainstem where there was little room for expansion. Only olfaction could really exploit the brain's processing and memory spaces. It needed to: It had to keep track of episodic airborne whiffs of odor, as the amphibian did; but, since it moved around, the early reptile also had to keep track of itself and the scents it was tracking on a map.

The part of the brain it used to do this is called the hippocampus (seahorse—it is thought by some to look like one). It is called this in mammals. In reptiles the area has a different name. We are going to use the word, hippocampus, for all species.

The olfactory cortex was the primary input to the hippocampus. Both areas were three layered paleocortex. The brain at this point in evolution was olfactory lobe on the outer surface and hippocampus on the inner.

The hippocampus was an area out of the direct flow of sensory data. Instead of being subject to a constant stream of sensory input, a memory trace could be perpetuated and processed without constant interruption. It converted the brain from a three- to a four-part device, now with an off-line map-space—a notepad.

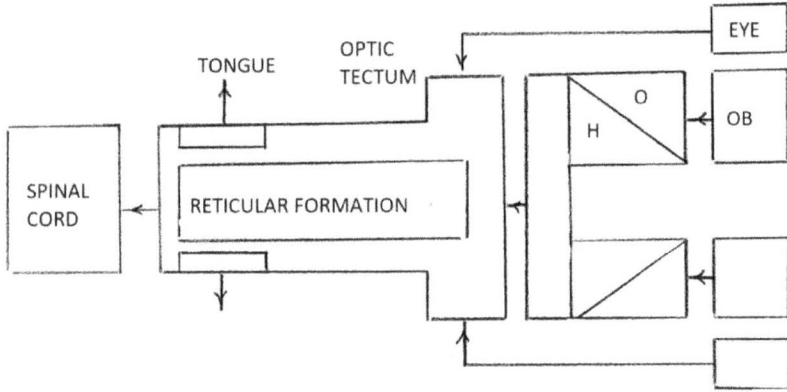

Figure 71 In the early reptilian brain the olfactory cortex (O) connected to the hippocampal cortex (H) which had a connection to the brainstem reticular formation through the diencephalic common base of the two hemispheres.

It was a map with synaptic memory: In the short term, like the olfactory bulb, it changed synaptic responses. In the medium-term, it changed the chemical composition of synapses. In the long term, it changed their size and number. It was a better memory.

(Amphibians have a simpler area like the hippocampus. It is discussed here because early reptiles would have had to use it more and would have been more driven to select for better function.)

Place Cells and Map

In the 1970's John O'Keefe recorded from rat hippocampal neurons and found cells that fired when it was in, or when it received signals from, a particular place. He called them place cells and thought they formed a diagram of the space around the animal.

In the ocean, everywhere is more-or-less the same, but on land it helps to have a map. If you are an early reptilian predator, it helps to have a map with your prey on it. Its hippocampus was a map of space and its food smells, a map of lunch.

This is the first time we have used the word map, and we will be using it repeatedly in the following chapters. As the brain evolved, it created maps or representations of the outer world and mapped sensory objects from that world onto them. There will be later maps for touch, hearing, and particularly vision. This one is literally a map: a model of the surrounding territory and would serve for decisions requiring knowledge of what was in that territory. (This is not the first map. The tectum is a multi-sensory map, but in neurology the word is usually used for a cortical map.)

A map is a memory, and the hippocampal map varies in size with memory requirements. In mammals, the relatively largest belong to animals like squirrels that must remember the locations of caches of nuts, and to the males of polygamous species that must remember the locations of females. Interestingly, humans have this male-female size difference, which leads to thoughts about men and women and marriage.

There is a hippocampal link between food and sex. To deal with its appetite for food, the animal uses a map. For another basic appetite, it uses the same map.

The Hippocampus and Space

How the hippocampus works in humans and mammals is incompletely understood. How it works in reptiles is less well understood. How it worked in primordial reptiles is unknowable, but it got inputs from the olfactory cortex and mapped them. In later species, it would evolve to map other sensory signals as well.

The circuit diagram of the mammalian hippocampus is complicated with multiple regions and loops. We do not know the details of the early reptilian version. We will assume it had output cells like the bulb, intrinsic cells like the bulb, oscillations like the bulb, three-layer cortex like the olfactory cortex, and recurrent pathways like the olfactory cortex.

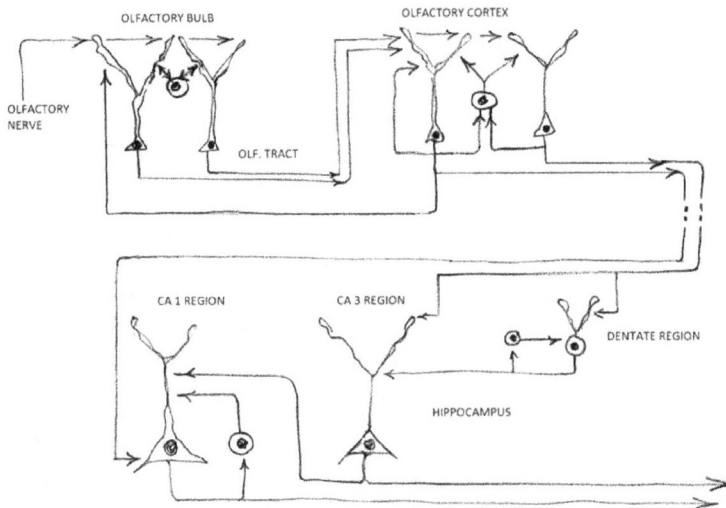

Figure 72 Circuit diagram from olfactory areas to a simplified mammalian hippocampus. It is similar to the bulb but more complex. The architecture and pathways are even more complicated than shown. We are not going to discuss the details or the names.

We do not know how it came to be. It is another case of "and now a miracle occurred".

We are not going to make a model of it but just discuss it in general terms. In general terms, it got signals from the olfactory cortex and located them in a map of space. What was the map like?

There are two types of maps: Subjective maps where objects are located egocentrically (near my right hand). Objective maps where objects are located in abstract, non-egocentric space (two blocks over and one block up from the park). The sensations from olfactory cortex arrive in egocentric orientation and are mapped into non-egocentric coordinates in the hippocampus.

What happens to the map if its map owner moves? An egocentric map must be re-drawn, but a non-egocentric one must only re-locate the mapping animal. O'Keefe and Nadel found cells that registered movements and head positions of the animal itself, using signals sent from the brainstem movement areas. The map is both a representation of the world and of the map carrier's movement through it.

The map's non-egocentric representation is suited to remembering territory. Egocentric memory is useless if you are at a different place on the map.

The map enlarges as its memory requires: It makes new cells. It grows with exercise. London cab drivers show increases after learning the London street-map.

This mapping will be essential to later cortical function. In later species, the hippocampus will be the gateway to long-term memory storage for all sensory modalities, both within itself and in neocortex. It can be thought of as coming at the end of sensory processing and the beginning of memory storage.

Its first job, though, was to map and remember objective space. Without it, that information does not register. For the tunicate, space does not exist. For all entities with a hippocampus, it not only exists, but forms a framework to organize sensory experience—the structure of the Umwelt.

Human Spatial Map Problems

People who cannot find their way around because of damage to this structure are rare. There are only a few reported cases of failure of the hippocampal topographic map, or atopognosia (a-no, topos-place, gnosis-knowledge).

One such woman was unable to find her way around from childhood but was otherwise neurologically normal. She took a bus to work and got off when she recognized a landmark city square. Once in the square she could recognize her office building. Any deviation from these egocentric landmarks got her hopelessly lost.

The magnetic resonance image (MRI) of her brain revealed a normal brain with two normal hippocampal areas, but an MRI of brain activity showed that they did not become activated when she tried to do a route-finding task. They were intact but disconnected from spatial mapping. This was thought to be a developmental connection failure.

The Amygdala and Motivation and Meaning

The brain is not just there to create pretty representations of the outside world. It is there to do something: to assign meaning, to make decisions, to direct action toward those representations.

Connected to the hippocampus is the almond-shaped area called the amygdala (almond in Greek). The amygdala assigns meaningful values to the objects represented in the hippocampus. It links them to the basic driving emotions. It acts as a map of appetite and motivation for action. It generates signals for approach or withdrawal.

It is called the organ of the five F's: fear, flight, fight, food, and—as the old neurology joke goes—making love; or, for those who dislike cute jokes, fornication. In both hippocampus and amygdala, feeding and sex and aggression are linked.

The amygdala makes the brain a five-part system.

The Spatial Map Model Animal

We could convert our model frog into a model early reptile by connecting olfactory cortex to hippocampus to amygdala. The model hippocampus could be designed to perpetuate its response for a time period after an input signal and so remember it. The side with the biggest input could activate the movement system to turn the model animal to that side and activate the amygdala to approach or withdraw.

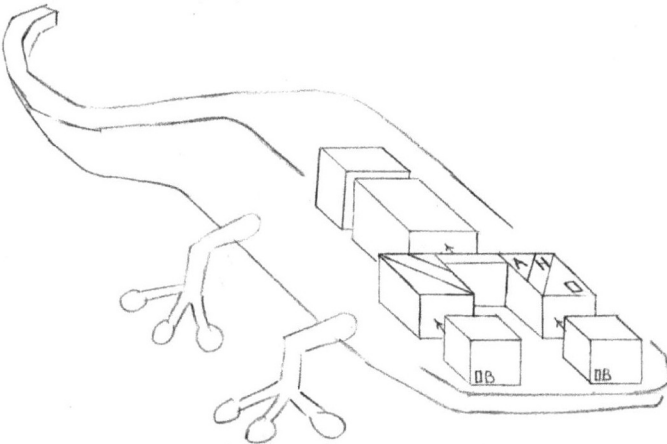

Figure 73 Model early reptile with olfactory bulb (OB) connected to olfactory cortex (O) which connected to hippocampus (H) and amygdala (A).

This would be an egocentric and quite limited spatial map. What if there were something in the way and it could not move toward the target?

It could put the directional bearing in hippocampal memory, and then move along until it could head toward the source on a new estimated bearing.

It might even, with two bearings, be able to locate the source. A sailor using two bearings and basic geometry can calculate the distance to another boat or a lighthouse on shore.

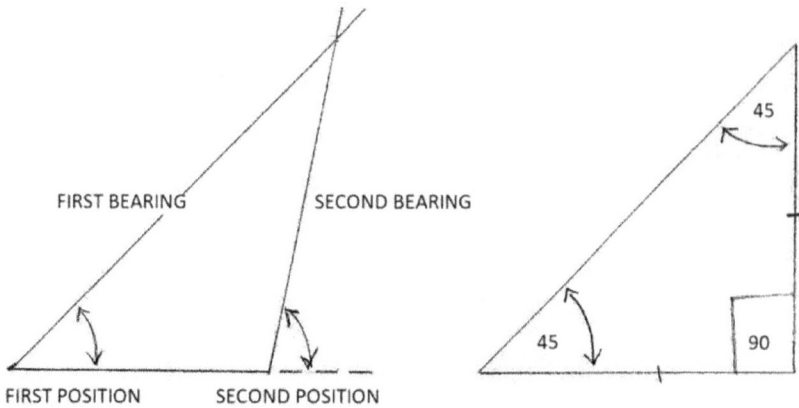

Figure 74 Two bearings and the distance from first to second positions allow the location of the source to be worked out geometrically. To the right is the simplest case of a right triangle where the sides adjacent to the right angle and marked with cross hatches are the same length.

This is not to suggest that this is what the hippocampus does but only that a map opens the possibility of working things out on the map, of playing with possibilities. An early reptile's hippocampus would not be able to do formal geometry, but it might be able to estimate bearings and locations—as we can, and without pulling out a protractor.

Such representations and processing go far beyond our models so far. We will not try to make a working model. We will leave that to the roomful of engineers with computers—and the room next door full of neurophysiologists with electrodes.

The Memory Model: Content Addressable Memory Systems

A map is a memory. The hippocampus is a memory system, but not like that of our best-known memory system, the computer. Computer memory uses location-addressing: Each memory location is a pigeon-hole with a numbered address. A central processor keeps track of the addresses in a register. This is not plausible in a biological system. As Tuevo Kohonen and Leon Cooper pointed out in the 1970's, the brain is more likely to locate by content or key pattern: A small input pattern, which could be a receptor vector, is matched to parts of stored patterns. When it finds the match, it retrieves the larger stored pattern.

The references in this book are content addressable.

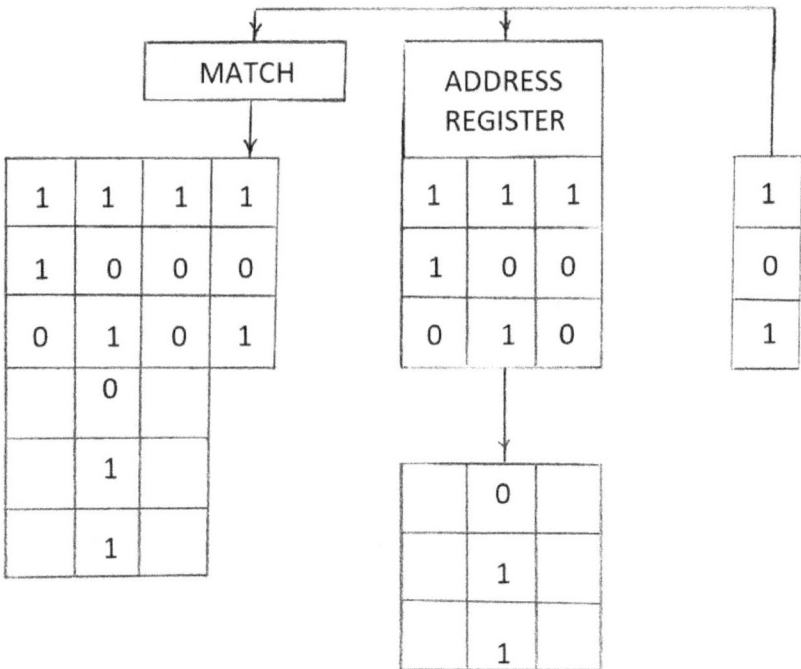

Figure 75 The location addressable memory on the right uses an address [101] to select a memory [011] in its address register. The content addressable memory on the left matches pattern [101] to part of a large memory pattern [101011].

The recurrent loop structure of the hippocampus is suited to content addressing. A small trigger pattern could generate a memorized pattern, and a recurrent loop could generate a larger pattern on the next pass or even loop again and again to generate pattern after pattern like a movie.

Could a connectionist memory act as a content addressable memory? How would such a memory work?

Let us propose a hypothesis: If the receptor patterns are different enough, then they will be able to generate and retrieve unique stored memory patterns using content addressing.

This seems to make sense. It might work. How could we know?

Let us translate the verbal theory into connectionist mathematical notation to make it model-able and test-able and see if it works. (This is the only place in the book where we do any mathematics, and that only multiplication. We refer to the result in later chapters.)

In the last chapter, memory was,

$$A'' = A + \Delta A' + \Delta A''$$

where,

$$\Delta A' = f'. r'$$

and

$$\Delta A'' = f''. r''$$

and if we now make a typical mathematical assumption that the initial A-term is small and can be ignored, then A'' becomes

$$A'' = \Delta A' + \Delta A''$$

$$= f'. r' + f''. r''.$$

and if the memory system gets another r' input and must retrieve the f' response, we can multiply through A'' with r' to get

$$fm = A''. r'$$

$$= (f'. r' + f''. r''). r'$$

$$= f'. r'. r' + f''. r''. r'$$

and now we can see that (and this is the tricky mathematical bit) if

$$r'. r' = 1$$

and

$$r''. r' = 0$$

then

$$fm = f'. 1 + f''. 0$$
$$= f'$$

and, similarly, if

$$r''. r'' = 1$$

then r'' will also produce its associated f''.

We can continue like this, feeding more distinct stimulus response signals into our model until it is a large content addressable memory,

$$A'''' = \Delta A' + \Delta A'' + \Delta A''' + ... \text{ and so on,}$$

in which any input will activate its unique memory trace and evoke its unique original output which could mean a tasty bug or a predator.

What about that tricky receptor property? If we translate it back into words, we end up where we started: If the receptor patterns are different enough (in a particular way), then their interaction outputs will be zero except for interactions with themselves and the system be able to retrieve stored memory patterns. By writing it out explicitly, however, we can see this property more clearly and try to implement it.

In the world of mathematics, this is easy to do. Looking at the usual receptor vectors

$$r' = [1,0]$$

and

$$r'' = [0,1]$$

one can see by inspection that a dot operator could have this property, and the dot operator of vector multiplication does. (The property is called orthogonality since the vectors are at right-angles on an x-y plot.)

In the real world, this would be possible if the receptor patterns were different enough. In the real world, such a memory would be plausible in hippocampus and cortex.

Human Hippocampal Damage and Episodic Memory

The hippocampus started as a spatial memory map but does more in the human brain. Damage to it affects memory in general.

The hippocampus in the early reptilian brain is where the olfactory signals go. In the human brain it is where all sensory signals go. Again, it can be thought of as the end of the sensory pathways and the beginning of the memory pathways. It is the major memory hub.

In 1953 a patient known as H.M. had both hippocampal regions removed to prevent epileptic seizures. He was evaluated afterward by his neurosurgeon, William Scoville, and psychologist, Brenda Milner, and found to have no ability to form new memories. He seemed normal in casual conversation but had no memory of what was said several minutes later. He recalled events prior to surgery normally but could learn nothing new. When he later looked at himself in a mirror, he found his aged face shocking. He was stuck in 1953 forever. (He had spatial problems as well, but these were minor in comparison.)

The memory traces for day-to-day events are stored either in the hippocampus or in the brain after processing in the hippocampus. If that structure is damaged on both sides, the ability to record long-term memory is lost. The old memory patterns set; no new patterns form; the memory crystallizes.

In bringing up human memory at this point, we are taking a giant step into the neurological future. Human memory is vastly powerful; it can store and retrieve a lifetime's memories. We are no longer talking about plodding neural networks but memory systems that can learn a picture or a poem at a glance and remember it forever.

It is called episodic memory: It remembers episodes in space and time. How it works is not well understood.

The hippocampus is essential for episodic but not all memory. H.M. could learn to learn new physical actions and modify his sensory responses. This uses simpler memory systems like those of the model olfactory bulb and is called procedural memory. Despite this, anyone meeting him who was not a psychologist would consider him to have total loss of memory.

How does memory for events in time relate to memory for space?

A spatial memory is an event in time. Perhaps the two are handled in the same way? Perhaps events are stored as sequences of spatial snapshots? Perhaps one snapshot evokes the next, and time is recorded like the stills of a movie?

This is speculation. How the hippocampus handles memory for time is not well understood. It is, however, the structure that evolved to register it. Without it, that information does not register: For the tunicate, time, like space, does not exist. For all entities with a hippocampus, they not only exist, but also form a framework to organize sensory experience, the underlying structure of the sensed world, the organization of the Umwelt.

John O'Keefe and Lyn Nadel thought this kind of mapping was a significant advance. In their book, The Hippocampus as a Cognitive Map, they argued that the hippocampus was different; that its mapping representation was more abstract and generalized; that its map memory was more precise and durable, and not prone to the blurring and generalization caused by overwriting memories. They thought that such mapping and memory would be the basis for later neocortical functions like thinking and language.

They thought that their findings about hippocampal space and time handling had been anticipated by the philosopher, Emmanuel Kant, who thought that space and time were fundamental human modes of registering experience. They gave him credit for the concept.

The paradigmatic disease of episodic memory is Korsakoff's syndrome usually from thiamine deficiency in chronic alcohol abuse and is similar to the H.M. state. The most common is Alzheimer's dementia with progressive loss of neurons, first in the olfactory cortex and then in the hippocampus with episodic memory trouble. It is a disease of global deterioration that later damages neocortex, but episodic memory loss is usually the earliest and the most devastating symptom.

Appetite Dysfunction in the Amygdala and the Kluver-Bucy Syndrome

Episodic memory dysfunction is at a higher level than anything we have discussed so far. Damage to the amygdala also causes high-level but even stranger dysfunction. Such animals cannot get their appetites right:

They are abnormally oral: They put all sorts of things in their mouths. They attempt to eat inappropriate things.

They are abnormally sexual: They attempt to mount frequently and to mount inappropriate sexual objects.

They are abnormally fearless: They are unable to recognize the emotional significance of objects like predators and show no fear of things that would normally frighten them.

They are abnormally unaggressive: They are docile.

They are abnormally incurious: They are uninterested in exploring the world around them.

Heinrich Kluver and Paul Bucy described this syndrome in lab animals in the 1930's. It was later found to occur in humans and has been reported in late-stage Alzheimer's dementia.

The Paleo-brain and Its Paleo-representations

The early reptilian brain had olfactory and hippocampal paleocortex. The spatial topographical patterns of the olfactory bulb are not found in either (in later testable animals). It is thought that this was because they are not sensory processing cortex. That the representations are not high fidelity and of the kind suitable for further sensory processing. That they are instead more abstract and more suitable for combining olfactory patterns with space and time patterns, for memory storage, and for going on to drive motor behavior.

They were the first abstract representations of things in the outside world. They are unknowns. They inhabited the first abstract representation of the outside world. It is an unknown.

We can make a limited model, but how close it is to the reality of the early reptile cannot be checked.

Content-Addressing Model

An animal smelling a particular place (r′) could send on a transformed version, or just a copy, to the hippocampal memory map which could have a memory of the area and the food it had found there earlier (h″), so that

$$hm = H. \, r'$$

$$= (\Delta H' + \Delta H'' + ...etc.). \, r'$$

$$= \Delta H'. \, r'$$

$$= h'$$

which could be sent on to the amygdala to be transformed into an approach signal (am)

$$am = Am. \, h'$$

which could be transformed into a brain output signal (x)

$$x = B. \, am$$

and sent to the reticular formation to direct movement to find food there now.

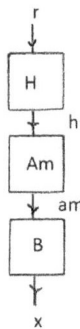

Figure 76 The hippocampus to amygdala to brain output model

Simple procedural memory like that of our olfactory model is reasonably well understood. Hippocampal episodic memory is not. There is, however, evidence that it uses similar synaptic changes, and that content-addressing is plausible.

Olfactory and Other Drives and Sigmund Freud

What would this be like? How could such an early reptile, dependent on its smell system and with little hearing or vision, sense and act?

It would get up in the morning, come out of its lair, and breathe in its environment. Smells would trigger spatial memories and memories of the big predator in the swamp to the left and the sexually active female by the stream to the right.

It might pick up a prey smell that would evoke a hippocampal representation and then an amygdala-to-motor representation to drive it up the smell gradient.

It might pick up the female's scent and be driven up that smell gradient.

It might see an insect and its tectum would bypass the slower smell system to trigger a tongue reflex.

It might be driven by hunger to explore its territory. It might use fixed smell landmarks: twenty feet out until it picked up that rotting-log smell and then bear right toward the place where there had been food smells before.

It might keep track of its peregrinations in a spatial map. It might lift its hind leg and leave scent markers to help it find its way back to its lair.

This world is closed to us, or at least to me. With my eyes closed, I can find my way to the coffee pot or the Christmas tree, but I follow a visual memory map. I cannot track up scent gradients and I am rarely motivated by them.

I am at times motivated by my amygdala, but not like the early reptile was driven by its. The reptilian brain model suggests Freud's hydraulic psychological theory with the amygdala standing in for the unconscious id and generating psychic pressures that push psychic energy down pathways to activate behaviors that satisfy primal motivations and reduce pressures. The closest human equivalent would be a soldier on point in advance-to-contact in a state of heightened awareness with all distance sensors scanning for potential threats; or a hunter, equally stimulus hungry but searching for potential rewards to trigger amygdalar responses.

The Processing Perspective

Olfactory input goes to olfactory percept representation, hippocampal world representation, and amygdalar appetite representation. These last two areas are in the limbic (border) region of the medial temporal lobe. There are two versions of each with connection pathways between the hemispheres.

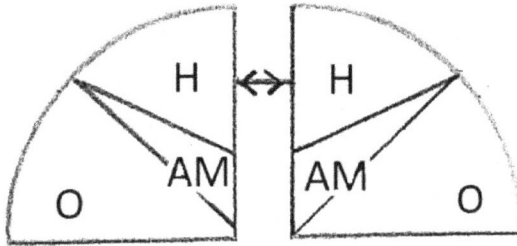

Figure 77 The hippocampus, amygdala, and olfactory representation areas with connection pathways to the opposite hemisphere.

These areas are representation systems with memory—blackboards that can be written on. What are the blackboards like?

The perceptual blackboard (O) is concerned with sensory object recognition and uses its memory to figure out what it is smelling. Networks can categorize and it can be thought of as a percept recognizer and categorizer: Animal or plant? Insect or alligator?

The amygdalar blackboard (AM) can be thought of as a situation categorizer: Prey or a predator? Reward or threat? Approach or avoid?

The hippocampal blackboard (H) can be thought of as a map and environment categorizer: What environment? Swamp map or river map?

The amygdala and hippocampus can be thought of as reference frames or context maps for the percept. In later versions, the context map will include a time context.

Context memory affects percept memory: A drinking song learned in a pub will be remembered better in that context. One learned by a campfire will be remembered better there and helped by the powerfully evocative smell of smoke.

The limbic lobe drove the simple behavior of early reptiles. Later animals will be driven more by their cerebral hemispheres.

For such a cerebral soldier, limbic context would form a situational framework for perception. The cortex would direct attention, perhaps to potential threats in forward tree-lines in advance-to-contact, and differently in defense, but humming in the background would be the limbic fight-or-flight aggressive mindset. Back at base when a pretty girl went by, an entirely different limbic mindset would frame the perception. The limbic lobe would modulate but not direct the response to the pretty girl—if the soldier's cortex were working.

The amygdala could take total control. I was once standing beside an open vehicle when there was a sudden loud noise behind me. I vividly remember being suddenly head down on the floor of the vehicle with no memory of moving.

In Erich Maria Remarque's, All Quiet on the Western Front, an old soldier on a peaceful day suddenly dives to the ground and an artillery shell explodes seconds later. The amygdala can learn dangerous sounds and react before conscious thought. This is called amygdalar hijacking.

In the cerebral cortex, Freud's volitional ego and conscience-driven superego hold the reins that control the primal id. From here on, we will be more concerned with cortical and less with limbic function which we will collapse into a limbic module (L). This is a made-up term. It is a theoretical module of all limbic activity that encompasses two maps in each hemispheric and the connections that allow them to work together, and the two hemispheric systems and the connections that allow them to work together.

Kant's World: The Early Reptilian World of Space and Time

The early reptilian amniote advanced onto the land and exploited it. The changes in its nervous system allowed it to find its way around on land, to communicate with its fellow creatures, and to hunt.

The animal was no longer just a stimulus response mechanism. The world of immediate response was now supplemented by a world of memorized signals that drove responses needing time to accomplish—an early kind of planning.

The framework for olfactory sensing was changed. Now there was an internal map of the outer world in the hippocampus. Smell signals could be located on the map and be searched for or avoided. They could be kept track of as the animal moved around.

The early reptile could use the map to keep track of prey in its hunting territory and be an active rather than a sit-and-wait predator. It could also use the map to keep track of its social, or at least its sexual, contacts.

This was a Kantian Umwelt with its memory map organized in patterns of space and time. It was not an egocentric but an objective map. It was not a sensory but a symbolic map. The animal itself was represented on that map, the first objective self-representation.

The Kantian map was linked to a Kluver-Bucian map of simple appetites: Food and sex; both linked to territory; both in turn linked to aggression.

The new maps directed action. Action in the old direct sense, but also a new kind of action intended to alter the behavior of other entities in the world—action to communicate intent. The communication was limited to sex and aggression, but sex and aggression are still major themes. We continue to communicate about them in daily life and in thousands of volumes on the bookshelves every year.

One form of communication, the maternal, was absent. The female early reptile laid her eggs and left them. There were no affectionate displays from infant to mother and back. If the mother later encountered her infants, she ate them.

In the early reptile we see echoes of ourselves. We still display. We still hunt. We still link the basic drives of sex and aggression: Rape is still a crime, and it is probably no accident that the most aggressive two-word phrase in English refers to our most intimate act.

We have made progress. We are nicer to our children.

PART III

MAMMALS AND CORTICAL MODELS

CHAPTER 8

ARCHAIC MAMMALS AND HEARING
AND CRAIK'S CORTICAL MAP WORLD

Ever since Texas, I have preferred to live in houses with balconies. My house in Florida sits in a grove of palm and pine trees. The light on the balcony is like that in a forest, dappled and changing. The background sound is similarly patterned and changing. The wind sighs, the leaves rustle, the palm fronds click.

Animals moving in the trees can be better heard than seen; and, at night, sound is the only way to detect them. This is the world the first mammals knew; they were night hunters.

There are anole lizards but most of the moving sounds are the squirrels. The palm fronds form three-dimensional expressways, and they rocket along them like Ferraris. The squirrels make the lizards seem about as lively as rocks, but evolution did not at first favor them. The early mammals were almost driven out of existence by the reptiles, and only chance allowed them to survive and later come to dominate the world.

We are going to discuss the innovation that saved them, good hearing. We will discuss sounds and the map of sound representations displayed in an internal model of the outer night.

Archaic Mammals

The early reptilian amniotes lasted for 145 million years and then a catastrophe called the Permian extinction destroyed 95 percent of animal species. The survivors' descendants included the later reptiles, among them the dinosaurs, and the early or archaic or synapsid mammals.

These early mammals were rat-like animals in the one-pound weight range. They were small predators with good teeth and big biting muscles.

Figure 78 The 4-inch-long early mammal, Morganucodon (205 MYA). The middle ear contained sound amplifying bones derived from the jaw joint. The cheekbone arch allowed the jaw to protrude up to attach to the biting muscles on the side of the skull.

They were warm-blooded. Their fossils have been found sleeping curled up with their young. Warm-blooded infants are less able to maintain their body temperature at night; they need contact with parents to survive night and childhood. Warm-bloodedness exerts evolutionary pressure for our kinds of behavior. It selects for interacting; it selects for communicating; it selects for a nervous system with areas devoted to these things.

These were not the keys to success at this time in the world's history. In the empty world after the Permian extinction, the small mammals lost the competition with the dinosaurs and retreated to a marginal niche of the dinosaur world. They stayed there for a long time, from 165 to 65 million years ago. The age of the dinosaur lasted a hundred million years.

What do you do if you live in occupied territory? What do you do if larger animals—monsters actually—are hunting you?

You hide. You hide where the monsters are not. Since the dinosaurs hunted during the day, a niche was left for the early mammals. They became night hunters. They selected for warm-bloodedness to allow them to be active at night. Then, when the dinosaurs could not warm up enough to sense or move effectively, the world was theirs.

The need to help the young stay warm led to selection for animals that took care of their young. Nursing the young was useful for animals that kept their young around for a while. The parent-infant interactions required more social behavior.

Hunting at night required better hearing. Hunting insects at night required better high frequency hearing. The mammals evolved a high frequency hearing system by modifying small jaw-related bones into sound-amplifying bones in the middle ear. They increased the size of the external ears, turning them into megaphones, and developed muscles to make them directional and able to focus on sounds.

Figure 79 A modern fennel fox with impressive auditory equipment. The biting muscles go down through the cheekbone arch to attach to the jaw.

Keeping track of the young at night in the dark was also possible by sound. The young made noises, and so did the parents. This could have led to problems during the day when the dinosaurs were up and hunting, but the high-pitched squeals of the mammals could not be heard by the low-frequency hearing systems of their enemies. The call of an infant separated from its mother could be heard only by her.

That separation call, according to Paul McLean, was the beginning of vocal communication and the basis for the mammalian social bond.

Paul McLean was a physiologist in the 1950's who developed the concept of the triune brain: The first part was the reptilian limbic brain. The second was the brain of early mammals which he thought was mostly auditory. The third was the large neocortical brain of later mammals.

The small neocortex of early mammals was the auditory processor that let them do well at night.

The Mammalian Auditory Brain

The mammal's habitat was tangled woodland: trees, fallen branches, rocks, gullies, streams, and ponds. It needed "...to navigate a junkyard in the dark...", as John McLoughlin put it in a book on early mammals. The distance sense that let it do that was hearing.

It needed to identify friend and foe in the dark. It needed auditory processing of what it was hearing.

The sounds it heard were episodic grunts and rustles in the undergrowth. It needed to remember them. It needed to make auditory representations and store them in memory.

It needed to remember where they came from. It needed an internal representation of the outside world with sounds kept track of in space and time. It needed an auditory world memory map.

This could not be done in the confines of the brainstem. Only the cortex would do. The mammal selected for cortical processing and memory. The brain increased in size, and between the olfactory and hippocampal paleocortex appeared auditory neocortex. Later neocortical vision and touch would follow.

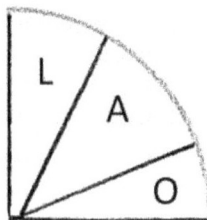

Figure 80 Auditory neocortex (A) between olfactory (O) and limbic (L) paleocortex which includes the hippocampus and amygdala.

Sound signals followed the old pathways from ear to both auditory nuclei and up both sides of the brainstem, but the major pathway was to the to the nucleus and tectum of the opposite side, and then to the thalamus and the auditory cortex of the temporal lobe.

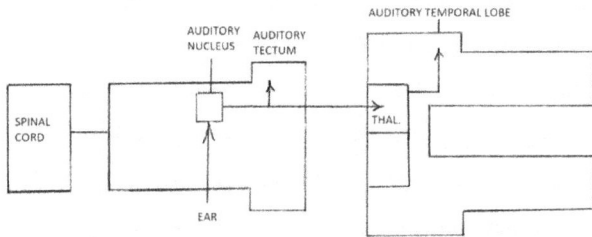

Figure 81 The early mammalian brain showing the major R ear to L auditory nucleus pathway going on to auditory tectum and auditory relay nucleus in the thalamus and auditory temporal lobe.

These pathways can be recorded from electrodes on the head. Sounds activate pathways and nuclei on both sides of the brainstem to evoke electrical responses called, not surprisingly, brainstem auditory evoked-responses.

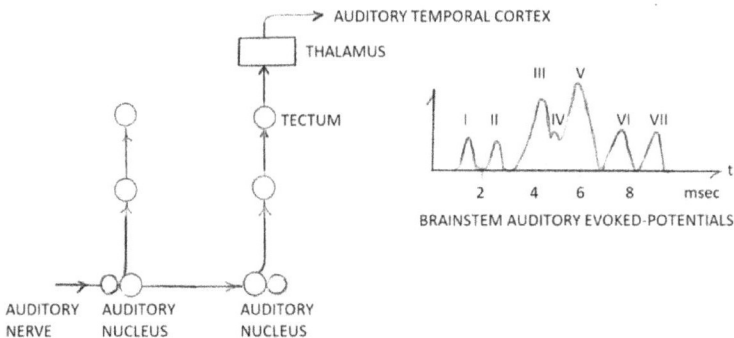

BRAINSTEM AUDITORY EVOKED-POTENTIALS

Figure 82 The auditory pathways where a sound stimulus travels through the auditory nerve to the auditory nucleus (beside it is the vestibular balance-related nucleus) where it splits to travel up the near and far side of the brainstem, relaying through brainstem nuclei to the auditory tectum, and opposite thalamus, and cortex. It evokes seven brainstem voltage responses in the first 10 milliseconds.

In electrical engineering this is called an impulse response: the stimulation of a system with an electrical impulse (or short sound stimulus) to see what it does. The brainstem does quite a bit. There are seven responses in the first ten milliseconds. The first is generated by the auditory nerve and the last two by the tectum.

Later responses can be recorded from opposite auditory neocortex at one hundred milliseconds and later. They are affected by thinking; the brainstem ones are not.

The new auditory neocortex had six layers: The three old paleocortex layers plus three new layers, intermingled but mostly on top.

Sound signals from thalamus went to relay cells in layer four, and then up to layers one and two to synapse on the dendritic trees of the pyramidal output neurons of layers five and six. Those in layer five went to brainstem and spinal cord as brain output. Those in layer six went back to the thalamus in a thalamocortical return loop: bottom up, top down.

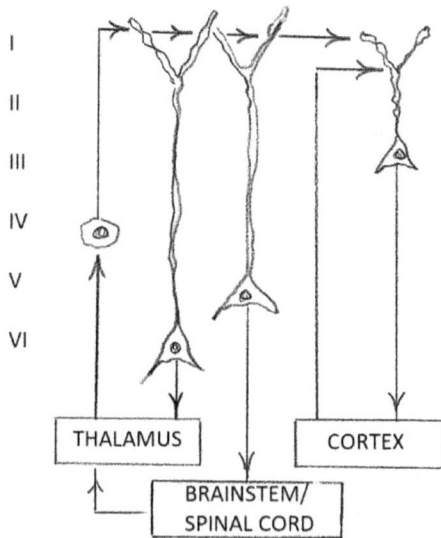

Figure 83 Neocortex (sensory and simplified): Inputs from thalamus go to layer IV relay cells and then to dendrites of pyramidal neurons of all layers. Large pyramidal output cells of layer V go to brainstem and spinal cord, and those of layer VI to thalamus. The smaller neocortical pyramidal cells in layers II and III get inputs from other areas of cortex as well and send their outputs mostly back to cortex.

The new neocortical pyramidal neurons were in layers two and three with their dendrites also in layers one and two. The cells were smaller and the cell-group modules smaller. Neocortex was finer grained cortex.

The inputs from the thalamus to neocortex were topographic patterns suitable for further processing. Neocortex was analytic cortex.

Upper-level neocortex, in addition to thalamic inputs, got inputs from other top neocortical layers, and these connected directly without the fourth layer relay. Neocortex was self-connected cortex.

Neocortical activity was an excitatory loop. It maintained its own patterns. Neocortex was self-driving cortex.

Sensory messages from the outer world entered only through the thalamus and only modified the constant interneuron activity. Neocortex talked to itself. It sometimes attended to the outer world.

The brain was now a six-part system that could do complex neocortical processing using short-term memory and store the results in long-term memory. The mammal was a neocortical animal.

It remained a paleocortical animal. The brain, like a power plant, must go on working during upgrades. Evolution cannot strip it down and redesign it. Changes are add-ons that must be compatible with what is already in place.

The new brain worked with the old brain. The brainstem tectum and the hippocampus connected to the thalamus. The three sensation-handling systems worked together.

The Auditory Cortical Map

What can we say about representation in the auditory system? How are auditory targets coded and represented?

The sound stimulus ($r(t)$) varies over and is a function of time. It is a more complex signal than a molecule lodging in a receptor.

The sound receptor is the cochlea, a coiled, snail-like membrane that, if unwound, would look like a triangular sheet. Different sized parts of the sheet vibrate at different frequencies ($f1$, $f2$, ...etc.), and the cochlea decomposes the signal into a set of sub-signals like do-re-me. Sub-signal $r1$ is a function of the controlling variables time and $f1$,

$$r(t) = r1(t, f1) + r2(t, f2) + ...etc.$$

The cochlea is a topographic map of sounds. That sound map is registered by neurons and sent to the auditory cortex where it is spread out in another topographic map of sound.

There is a mathematical technique called the Fourier transform that works like this. It decomposes a time signal into a series of sine waves of different frequencies. The size of each indicates how much it contributes to the signal. The Fourier transform frequency representation of the time signal is a graph of the amplitudes of the various sine waves. It could be written as a vector of frequency amplitudes, a distributed representation.

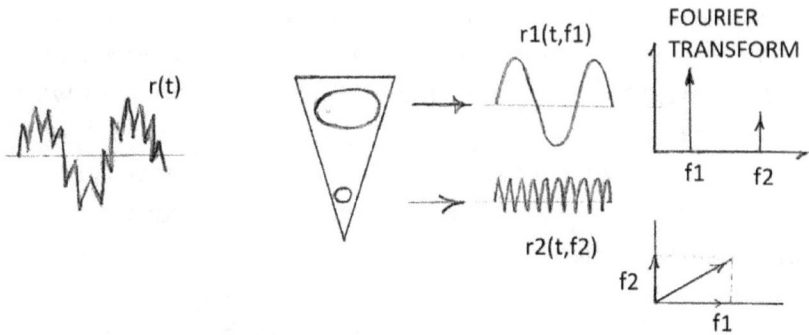

Figure 84 An input soundwave r(t) at left is decomposed by the schematic unrolled cochlea into an r1 wave with a low frequency of f1 and an r2 wave at a higher frequency of f2 with a smaller vibrating area. The Fourier transform representation has two components, the larger amplitude f1 and smaller f2, and could be represented as the vector at lower right.

The essential thing to understand about the Fourier representation is that it is a transformed but intact version of the original signal. It contains all the same information and can be transformed back into the original time series without any loss of fidelity. A Fourier transform with enough frequency components can represent any time signal. A list of frequency amplitude numbers could be fed into a processor and out the other end could come Frank Sinatra singing a smoky ballad.

That our topographical sound map has distributed sound representations is demonstrated by our ability to explore them. We can tune into spots on the map. We can attend to the violins up high, or the base fiddle down low, or Frank.

Craik's Internal Representation Hypothesis and Predictive Coding

We know that the auditory cortex has auditory representations, but we cannot measure them. We do not know what they are like, although some suspect they are Fourier transforms. They must be patterns in the brain kept in register with a map of exterior space—something over there. They must be modifiable to allow location changes and changing perceptions of what is making them—that something-over-there is moving this way and is one-of-those.

The coding of objects in the outside world into our brain maps is the key to how our brains work. We do not know how it is done. This inability to measure representations, or find the code, has been a source of frustration for many. One of the most talented of the mid-twentieth century researchers, Karl Lashley, spent years looking. He called the representation, the engram, and he summed up his work in a 1950 paper called "In Search of the Engram" in which he said that he had not found it.

Lashley's work could be considered a failure, but it did lead to a conceptual breakthrough. He had been looking for discrete engrams in specific locations. When he could not find them, he hypothesized that they were not localized but distributed. This idea has stood the test of time. The distributed representation models we have been using are consistent with Lashley's hypothesis. They have still not been measured.

A contemporary of Lashley's named Kenneth Craik developed another hypothesis about representations. He thought the mammalian brain contained "a working physical model of reality", in which sensations from objects in the outer world were related to representations of those objects in the model. The model had a "relation-structure" to the outer world, by which, he said, "I do not mean some obscure physical entity which attends the model, but the fact that it is a working physical model which works in the same way as the process it parallels."

This is a departure from the perceptual processing systems we have discussed so far. It is far from the commonly held idea of percepts as little images of the outside world. It is counter-intuitive, just as Darwinian variation and selection are counter-intuitive to deterministic design. This is nowadays called a mental model system: the brain contains a model of the world and objects in it and interprets stimuli as relating to the representations in the model.

It is also called a generative perceptual system: The brain generates its own perceptual representations and uses sensory inputs only for guidance.

It is not a complete departure from the perceptual processor concept. Those sensory processors recognized patterns in the blooming buzzing sensory confusion, stored them in memory, and later recognized sensory inputs as relating to them and retrieved the correct memory trace. Such stored patterns are internal models or could evolve into them. Perceptual processing systems could become mental model processing systems.

Built into Craik's hypothesis is the concept of functionalism. A brain model obviously cannot work like an object in the real world, but it can mimic its workings so closely that it can function "in the same way as the process it parallels".

As such, it can be used to anticipate what will happen next. Another version of the hypothesis goes by the name of predictive coding: The system is thought to not only generate representations but also to predict future changes to them. It compares the actual sensory inputs to its predictions and then corrects. This would save processing work for inputs that do not change much, and much of what we sense does not change from moment to moment.

We are not going to go into the pros and cons, but simply adopt mental model perception as a working hypothesis. We will assume that the Craik internal model can:

> recognize a sound with a pattern-recognizing perceptual processor,
> assign it to an internal representation,
> display that representation in a spatial model of the outer world,
> manipulate the representation,

predict,
compare new inputs,
update the model.

The hypothesis is consistent with the overall theme of the brain being a model of the outer world.

It is consistent with the idea of it learning and predicting. The learned receptor-set responses of the model fish can be thought of as a prediction of what is likely to be swimming around in the future.

Learning itself can be thought of as a prediction that the future will resemble the past. There is no point in learning if it does not. Modelling is similarly pointless if there is nothing consistent to model.

The Craik model seems a plausible way for a nervous system with learning to learn to function.

A Speculative Change in the Meaning of Meaning

Complex perceptual processing has consequences. The most important is a new way of assigning meaning: Meaning is the matching of a sound to an internal symbol. The perceptual apparatus is now representing meaning.

I have flagged this idea as a speculation. It is not in Craik's book but does seem to follow from the mental model theory.

I must qualify it as well: It is unlikely that this is the final word on the subject of meaning. I should say instead that some of meaning is the matching of a sound or other stimulus to the memory trace of its internal representation.

The Internal Representation Map Model

Sounds would be recognized as percept representations in a percept memory map: tree noises like rustling, insect noises like buzzing, and predator noises like roaring. The percept representation symbols would be displayed in an internal world map, a model like the hippocampal one, but in the cortex.

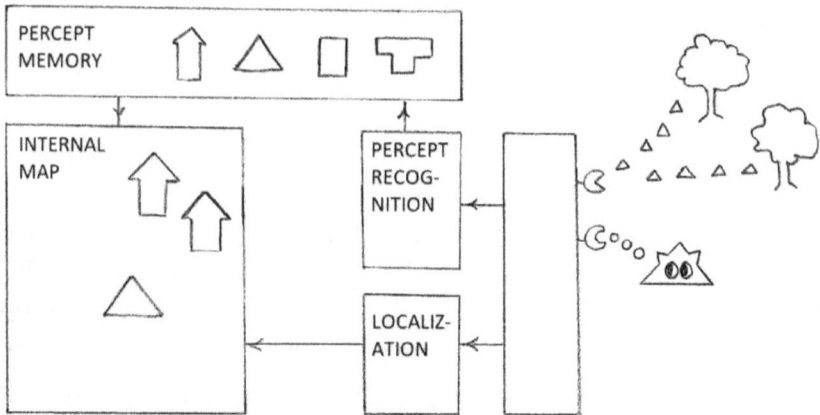

Figure 85 Internal Auditory Map: The trees and the creature in the real world generate sound stimuli that are registered and compared to percept memories and then displayed in an internal map of external auditory space. A localization module would assign spatial locations.

To put the representation in the right place on the map would require a location signal. A right-to-left volume comparison would allow assignment to the proper side. A front-to-back localization would be more difficult. We could assume our animal has four ears—this is modeling, after all—but directional ear movements and time-of-arrival differences would allow a more mammalian method.

The hippocampus could keep track of locations, but cells like its place cells are found in neocortical areas as well. We will treat localization as a single module, but it could be distributed, and in both areas.

These nocturnal mammals inhabited a quite spectacular sound landscape. We do not. We are visual mammals. We do not think about sound much. It is a background sense. We notice if there is a loud noise. We notice if something changes. We attend mostly to speech.

The Algebraic Brain Model

We are not going to build a working model of auditory cortex, but we will assume that a sound representation (r') would trigger a memory representation (fm) in the usual way In a percept memory map.

We will simplify the notation by writing the percept memory (A) as containing only the ΔA memories, so that

fm= A. r'

= (ΔA' + ΔA'' +...etc.). r'

= ΔA'. r'

= f'

but Craik's hypothesis says that the transform would take place in the brain with the sound evoking a memory representation that is part of the brain activity vector (b).

We have treated the brain as a sequence of transforms of sensory inputs to arrive at motor outputs, which is to say a stimulus-response system. In the last chapter, we skipped over the possibility that the hippocampus might require more. Now we need a model with a separate pathway of brain activity.

The simplest such model is

x= A. r + B. b

which says that the brain output vector (x) is determined by the sensory input vector (r) and the brain activity vector (b), which would include the hippocampus and amygdala.

The input r-vector also changes the b-vector into the next b-vector (b+1)

b+1 = C. r + D. b

and it is also changed by its own ongoing activity (D. b).

Another input vector (r') would be transformed by memory into ΔA'. r' and ΔC'. r' and affect both equations.

This might look a little intimidating, but these equations just state the obvious: What we do is determined by what is happening around us and what is going on in our brains. What is going on in our brains is determined by the same two things.

These are the equations of the state systems model used in electrical engineering for signals processing and control systems. The fundamental concept is that such a system is completely determined by its inputs and its internal state. They are the only controlling variables.

The math can be difficult, but we are not going to do any. The model is unfinished, and we are not going to finish it. Writing the equations down is just the beginning. The hard part is writing the transforms and in this case that means figuring out how the brain works. We will sidestep this problem. We will treat the equations as a suggestive mental model and not try to turn them into a working model.

If we are going to treat the brain as two symbolic equations, we must consider how the symbols relate and how the controlling variables control.

If brain activity is a big vector, a mathematical function, what is it a function of?

Of its inputs and itself. Of time, of synaptic and neuronal and sensory activity, as we said before, and now of what is going on in Craik's internal model.

One of the things going on is learning. Where does learning occur?

In the transform matrices. They would do the changing and learning, but at different rates. Matrices A and C would change how they process inputs slowly. Matrices B and D would change rapidly—at the speed of thought.

Another is processing. We have described the brain as a sequence of processing transforms, but we show only one.

Any matrix (B) can be decomposed into multiple matrix stages which can be multiplied together as

$$B= B1.B2.B3,$$

and each sub-matrix could be a separate processing stage such as hippocampus and amygdala in the previous chapters.

Another is sensory input. It obviously must affect brain activity, but does it directly affect brain output as the first equation says?

Craik's hypothesis seems to say, no, that instead, input is translated into the internal model and only the internal model determines brain output. So, the first equation should be

$$x= B. b.$$

This could be correct but there are two reasons to think it is wrong: The first is that the evolving brain has no way of getting rid of old ways of doing things. It can add on but not erase, so the old stimulus-response processors never go away. We still have knee jerks.

The second is subjective experience. The most telling is the sexual: The flood of sensory input wipes out conscious thought. The brain becomes a stimulus-response organ in a thoughtless state the French call the little death.

Similarly, a skier on a downhill run, or any totally involved athlete, can turn everything else off and respond only to the hill or the game. They can turn off the b-vector and attend to the r-vector and pure stimulus-response. They can then go home, put up their feet, turn off the r-vector, and concentrate on the b-vector, the memories of the ski run or even how the brain works. This does create a problem: the model, as written, has no way to do this switching. We will get back to this later.

Another issue is linearity. All the algebra we have done so far is linear. Linear systems are easy systems where two times the input results in two times the output and two simultaneous inputs produce the same two outputs. Engineers prefer to work with such systems. Is the brain this simple?

The brain is not linear. Even its sensory nerves handle input in a non-linear way: Too much touch or sound or light and they stop working. We can, however, adopt a standard engineering approach and treat it as linear to a first approximation for a limited range of inputs.

A final issue is (b+1) which indicates that the model works from a clock pulse with discrete time intervals. The brain does not, but models do, so this is a reasonable modelling approximation.

Mental Model System Problems

The Craik model is a more complicated perceptual processor. What sort of trouble can it get into? Certainly, more complicated trouble.

There can be failures to interpret and assign. Damage to the sound recognition system of the left hemisphere causes inability to understand speech. Damage to the non-verbal sound recognition system of the right hemisphere causes inability to interpret sound:

You cannot tell a bus from a waterfall or a lion's roar. You cannot appreciate music; it is just noise. You cannot assign meaning.

These are complex perceptual pattern recognition failures. They have a special place in the hearts (and minds) of neurologists and their own special name. They are called agnosias (a-without, gnosia-knowledge). Hans Lissauer called them "sensing stripped of meaning"; or, more dramatically, "soul blindness" (Seelenblindheit). The patient can sense but not interpret. There are also agnosias of vision and touch.

There can be partial failures: bad pattern recognition or wrong meaning. Crackling cellophane can sound like fire. Wind, like a baby crying. These could be called dysgnosias (dys-bad), but neurologists prefer the term, illusions.

There can be things on the sound map that should not be there, such as representations with no sensory inputs. These could be called, pseudognosias (pseudes-false), but neurologists prefer the term, hallucinations.

There could be sounds that should be on the map but fail to make it through the system to get there. This could be called asthenognosia (asthenia-without strength), but neurologists do not make such a diagnosis. There is a condition called auditory neglect that may be a candidate: A lesion on one side of the brain causes the patient to neglect sounds coming from the opposite side of space. The patient can hear normally, and perceive sounds from that side if directed to, but does not normally do so. Do the signals fade away before they reach the map?

Map Mismatch

What if there is a change in the sound landscape? What if a signal changes in character or a new signal appears? What if it is a predator signal?

We can record this happening in the electrical responses of the cortical processors. If a monotonous series of identical beeps is played, the evoked response in the temporal cortex to each beep is a negative waveform at 100 milliseconds (the N100) and a positive one at 200 milliseconds (the P200). The response is always the same.

If something different happens, a boop instead of a beep, the N100 stays the same but a larger and later positive wave appears at 300 milliseconds, the P300. These are known by the inelegant term, beep-boop studies.

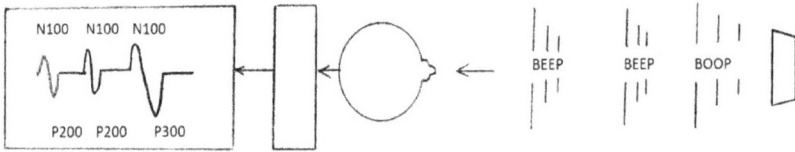

Figure 86 Beep-Boop Study: The speaker to the far right generates the tone sequence beep-beep-boop. The recording system in the box amplifies the brain signals and displays an N100-P200 response to the first and second beeps, and an N100-P300 to the boop. (Negative is up in neurophysiology for historical reasons. Multiple responses have to be added to see the tiny waves.)

The P300 is a memory comparison made visible: The usual tone must be saved in memory, compared to the unusual tone, and the difference noted. It indicates a sensory-signal-to-map mismatch and alerts the animal to re-write the map. Such a mismatch detector would be useful in a predictive coding system.

We can only tell beeps from boops if they are spaced at intervals of 300 or more milliseconds. A third of a second is as fast as we can make simple auditory discriminations.

Sensory Deprivation Hallucinations

What would happen to such a representation system if its sensory inputs were cut off? Its internal activity would continue. Would it keep on generating representations?

This happens. Patients with hearing loss can hallucinate sounds, often music. One of mine hears John Brown's Body playing over and over. Patients with visual or tactile system damage will hallucinate in those modalities. A person with a completely normal nervous system, if put in a sensory deprivation chamber, will within hours hallucinate in all three.

The Processing Perspective and Brain Sensory Flow Model

We will assume that sound representations are vectors with multiple frequency components, possibly Fourier transform vectors.

The neocortex model assumes only two processing streams. Brain output (x) is the transformed receptor input vector (A. r) combined with the transformed brain activity vector (B. b).

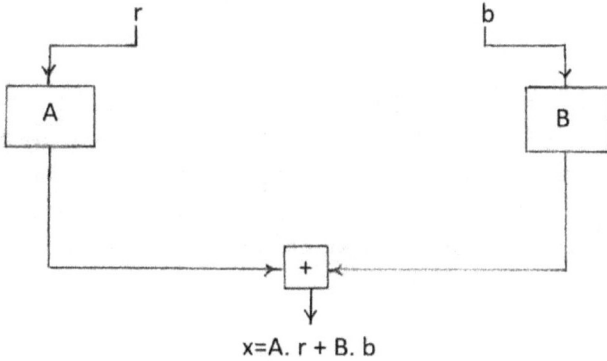

$$x = A. r + B. b$$

Figure 87 Brain output activity vector equation block diagram.

The brain activity vector (b) is changed by both input and by itself into (b+1).

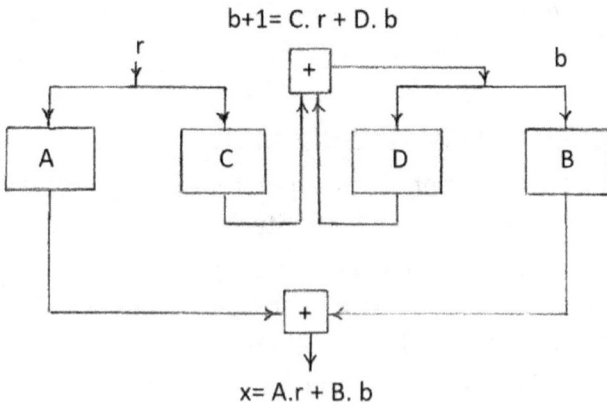

$$b+1 = C. r + D. b$$

$$x = A.r + B. b$$

Figure 88 Brain equation block diagram.

The processing of (b) into (b+1) is complete by the time the next receptor input enters the system. The model has processed its inputs and its internal content. Is it thinking?

Rather than trying to keep track of four processing streams, we will consider sensory flow only in one pathway. We will be most interested in the C and B pathways but will think of the four streams as processing in much the same way and interacting.

Sensory processing for all modalities, according to Pandy and Seltzer, can be thought of as having three general stages. For auditory input, the initial auditory processing stage (A for auditory) is followed by a stage (I) with more abstract representations that allow connections to other pathways and even other sensory modalities, and then a stage (II) with higher-level connections such as to the limbic area (L).

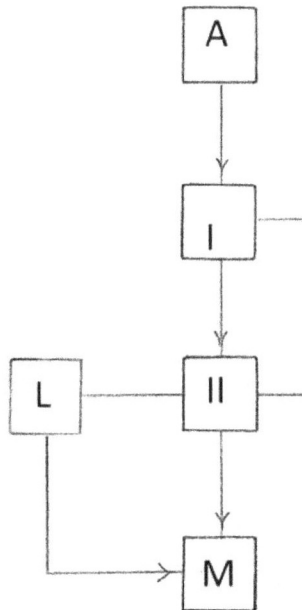

Figure 89 Auditory sensory processing flow sub-divided into three stages with the second connecting to the other processing pathways and even other sensory modalities, and the last connecting to higher processing areas such as the limbic (L) area and possibly the internal auditory map (M). (After Pandya and Seltzer, 1982. The M-map is not part of their model.)

The percept processing streams would go on to even higher processing levels. One of these would be the internal auditory percept-to-percept-memory-map (M) where

$$fm = M. r'$$

$$= (\Delta M' + \Delta M'' + ...etc.). r'$$

$$= f'.$$

The model processor could suffer percept recognition trouble and make a mistake in retrieval like

$$fm = M. r'$$

$$= (\Delta M' + \Delta M'' + ...etc.). r'$$

$$= f''.$$

This would be something like an illusion, and an absent response would be something like agnosia.

A pathway set to expect an (f') that got an (f'') instead could also function as a beep-boop mismatch detector.

Craik's World: The Internal World Model

The archaic mammalian brain moved auditory processing up into the neocortex where it matched auditory patterns to representations in an inner model of the outside world. This is complex perceptual processing. This system could be considered a sixth functional component of the nervous system and a new way of perceiving the world.

This inner world was necessitated by the mammals' situation of being hunted by monsters in the outer one. John McLoughlin wrote, "...symbolic thought was born in the terrible Mesozoic night". The inner world saved us from the monsters, but he wondered if "...deep, deep down...in some dreadful corner, the Arcosaurs still stride about like great mad birds from whose glittering, stony eyes we must at all costs stay hidden."

There are odd and counter-intuitive things about perception. It should be clear by now that perception is indirect, and that representations are not little imprints of objects in the world.

Although smell signals are actual pieces of the stimulus object that attach to nervous system receptors, their representations are quite different. The other modalities are less direct: sound signals are air pressure disturbances with no direct connection, and the visual signals of the next chapter are colored shadows—or rather shadows to which the nervous system assigns colors.

There are odd and counter-intuitive things about mental model perception. Complex perceptual processors match internal percept memories with signals from the receptor set and assign probable representations. What is perceived is meaning.

The meaning can be wrong. These assignments are based on limited perceptual data. The receptors register only part of the input signal and may distort what they do register. Perception is not only subject to their Umwelt limitations but also to the limitations of the representation systems themselves. Such systems can create problems, as we have noted before, and complex systems can create complex ones like illusions and hallucinations.

There are odd and counter-intuitive things about cortical representations. They are abstractions of the original stimulus. They are generalizations of repeated sensory presentations. They are constructions linked to sensory data—more like descriptions than sound recordings. They are brain words or symbols, and so determined by the brain's structure and ways of representing. They can be wrong or incomplete.

They have a peculiar status: They are the intermediate entities between objects in the world and us. They are the inhabitants of Craik's inner model of the outer world. They are symbols in a symbol space. They have never been measured.

There are odd and counter-intuitive things about the representation of the world they occupy. It is a perceptual theatre representing the outside world in space, but an electric impulse-train theatre. A receptor for the structure of the outer world, but one we cannot measure.

It is also a representation in time. It can operate, as Craik wrote, as "...a kind of distance-receptor in time, which enables organisms to adapt themselves to situations which are about to arise". It can project into and anticipate the future—at least when and where a baseball will land. It can use prediction coding.

How did this perceptual and representational revolution come about?

I am going to default to the cartoon scientist's position and say that another miracle occurred. I am not going to try to explain how it came to be, but I am going to say that it changed the nature of the inner world. It is self-generated. It is a construct.

It changed the nature of perception. The apparent imprinting of the outside world on the senses is an illusion. Perception is a fabrication projected out upon the outer world. The physiologist, Antonio Llineas, said: "Reality is a dream modified by sensory input."

CHAPTER 9

MODERN MAMMALS AND VISION AND PERCEPTUAL SENSE ORGANS

Once upon a time, I came upon a majestic tree shimmering in a balmy breeze with its leafy reflection dancing upon the coruscating surface of an azure pond.

I said," Wow. Look at that tree. It looks like a Watteau painting."

My partner looked up from his putter, gazed across the water obstacle, and shrugged. "Mike. It's just a tree."

We modern mammals, and particularly we primates, live in a glorious visual world—some more so than others. It is not what it seems. It is another fabrication. It is a construct of the neocortex that came into being when the mammals moved out of the dark.

We will discuss how the neocortex did that.

Modern Mammals

For a hundred million years, the mammals hid in the dark. Then, 65 million years ago, a meteor struck the Yucatan peninsula and caused a global winter known as the Cretaceous-Tertiary Event. In the mass extinction that followed, the large dinosaurs died off and the little mammals inherited the earth.

The mammalian species expanded to fill the empty ecological niches. They got bigger, and they got brainier. The mammalian brain, which had been stable for 100 million years, increased eight times in size. Not every animal changed. The smaller brained animals remained in the small-brained Mesozoic niches where they continue today and threw off brainier cousins to do other things. The range of mammalian brain size increased.

Around 55 million years ago, there were squirrel-like mammals, called insectivores, insect eaters, who took to living in the trees of tropical rain forests. They became the early primates or prosimians.

Figure 90 A modern primate Tarsier with big eyes and opposable thumbs.

Living in a tree requires good vision: You must see well to catch the next branch. You must see well if your prey is a flying insect.

Evolution selected for better visual processing. The retinas evolved areas of densely packed photoreceptors for detailed vision. The eyes moved forward. The visual fields of the two eyes overlapped and the images had to be put together in register. This allowed depth perception but took increased processing. This could only be done in the cortex. The primate became a fully cortical animal with all sensory modalities handled in the brain.

Another food available in trees is fruit, and this drove selection for color vision. Around 40 million years ago, repeated duplication of the retinal receptor gene led to the three receptor cells with different light frequency responses and the color vision system we use today— to see yellow bananas against green leaves among other things.

Another factor driving brain development was social behavior. The primates were group animals and selected for brain areas devoted to face recognition.

The tree environment also selected for visual-motor control. The prosimian developed an opposable thumb for grasping tree branches. This worked better if the visual areas could rapidly affect hand control, and so drove selection for cortical control of motor action. Fast tracts going directly from cortex to spinal cord began to dominate motor control. The reticular formation decreased in importance. As J.M. Allman put it, "The action center of the central nervous system moved up into the brain."

More cortical processing meant more cortex. The primate brain was now a big and very fancy piece of equipment. It was fifty times the size of a non-mammalian brain. It was the center of the central nervous system. It was the processor of sensory inputs and the controller of motor outputs. It was in charge.

It had superseded the brainstem but was still dependent on it. The upper reticular formation contained sets of neurons that projected up to activate the brain. This is called the ascending reticular activating system and is vitally important. It controls wakefulness. It is the light switch that turns your brain on every morning; if you damage it, you end up in coma.

Vision

Vision is the major sense in the primate and human brain. It is split into two halves. Light from the right side of visual space goes through the pupils to the left halves of both retinas and then to the left side of the brain. Visual processing is almost totally neocortical. The optic tectum is more concerned with eye movements.

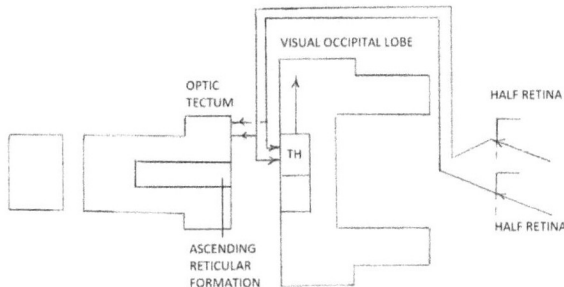

Figure 91 The visual signals from R side of visual space go to the L half of each retina and the optic nerve pathways join and travel back to the L thalamus (TH) and then to the L occipital visual cortex.

The two eyes see two versions of the same scene, and these have to be combined. The pathways from each left half-retina go to the left thalamus and then to the left occipital area where each half-retina displays its signals in adjacent strips of receiving cortex.

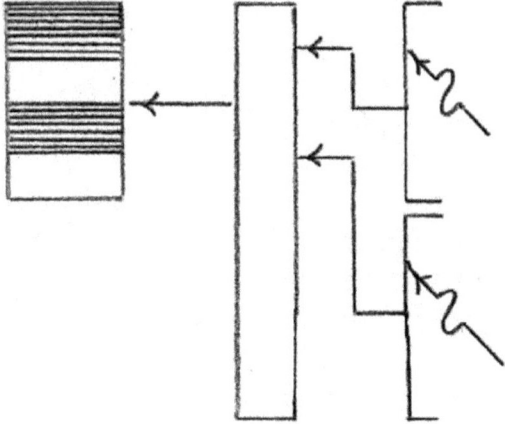

Figure 92 The squiggly arrow light signals from the R side of visual space going to the L halves of the retinas, L thalamus, and then L cortex where they register in strips of neurons for each eye with the L eye strips shown as hatched.

Real cortical hemifield strips look more like this.

Figure 93 Cortical hemifield strips with one eye in black and one in white.

The visual areas at the back of the brain are more than half of the neocortex. We have visual brains.

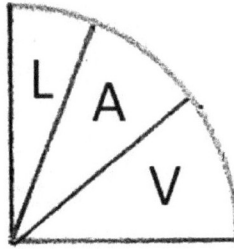

Figure 94 Visual cortex (V) is more than half of the neocortex. The olfactory area is not shown. It is now relatively small, and the expansion of the neocortex has pushed it under and over toward the limbic lobe (L) and its hippocampus.

Visual Processing and Edge Detectors

Vision is not what it seems. It is more of a telegraph message than a picture.

The visual world that is displayed on the retina activates photoreceptor proteins in neurons of visual cells. The responses are processed extensively in the retina, but we are going to skip the details and only discuss the output cells of the retinal modules, the ganglion cells. David Hubel and Torsten Weisel, working in the 1960's with cats, found some that were excited by small spots of light at particular locations and inhibited by light striking in rings around the center spots. They called them center-surround cells. These neurons are light-spot detectors.

The light-spot-detector representation could be modeled as a two-dimensional grid of ones and zeros and written as a two-dimensional vector. It could be written as one-dimensional vector or a long binary number if the cortex kept track of the grid localizations.

Damage to the spot detectors causes blind spots, from small holes to total blackness. Such deficits could be modelled as areas of unchanging zeroes.

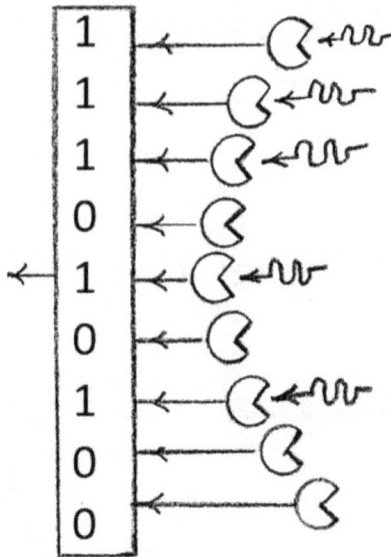

Figure 95 A schematic retina with light striking the photoreceptors of some of the neurons and the output of the ganglion cell responses modelled as a vector of ones and zeroes.

Hubel and Weisel went on to investigate the visual cortex and found processing to be complicated. There are more than forty visual processing areas with pathways operating in series and in parallel.

They found neurons there that sent suppressive impulses to the neurons around them. They called this lateral inhibition and thought it reduced activity from areas of uniform illumination and increased activity from the areas of differences in illumination. These neurons were edge-detectors. They thought this pathway recognized lines and shapes and called it the V1 line pathway.

There was a parallel pathway with neurons that recognized color. They called it the V1 blob pathway because of its blobby balls of neurons. These two pathways went on to the inferior temporal lobe.

There was another pathway (MT) concerned with movement detection and eye-directed-movement. It went to the medial temporal lobe.

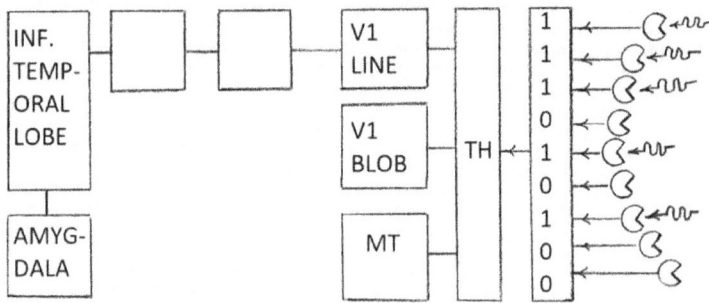

Figure 96 Block diagram of some of the forty stages of the visual system. The retinal representation is sent from the thalamus (TH) to the V1 line and V1 blob pathways that go to inferior temporal cortex, and to the movement detection pathway (MT) that goes to medial temporal cortex.

Further along the V1 line pathway, there were neurons that detected lines at specific orientations. Three spot-responses in a row constituted a line at a particular angle and triggered a cortical module registering that line. Adjacent modules responded to lines at different angles. These were oriented-line-detectors.

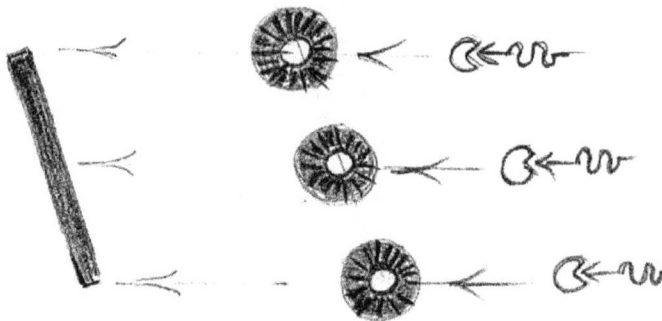

Figure 97 Light strikes three center-surround retinal cells in a line and a later set of nerve cells interpret this as a line or edge at a particular angle. (After a diagram in Mountcastle, 1974.)

David Marr's first hypothesis of visual processing was that the initial task was to make a two-dimensional edge outline or "primal sketch". Such a primal sketch could be constructed with oriented-line segments.

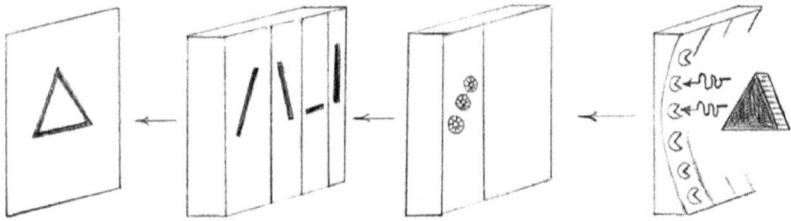

Figure 98 Marr's Primal Sketch Hypothesis: Light from a triangular object is registered by retinal receptors, which activate sets of line detecting center-surround cells, which go to a bank of an oriented line recognizers, and three such oriented line segments activate a primal sketch of the triangle.

Marr's second hypothesis was the percept-to-percept-memory pattern-recognition system of the last chapter, but for vision rather than sound. A primal sketch is a shape and a percept pattern to be recognized.

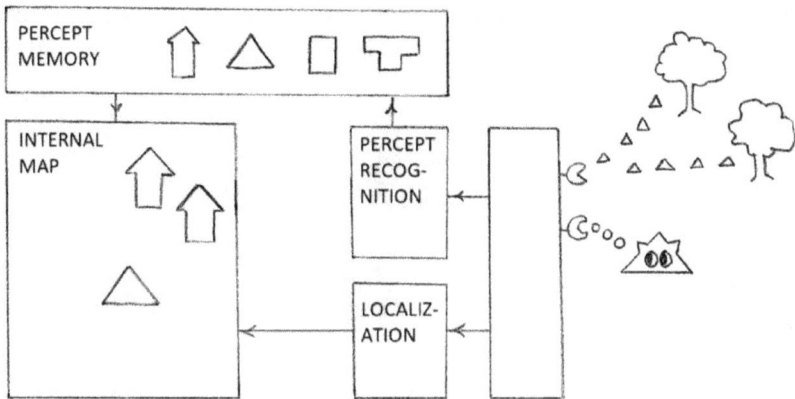

Figure 99 Marr's Percept-to-Percept-Memory Recognition System Hypothesis: This is the same diagram as the auditory version of the last chapter, but the stimulus shapes now represent light signals rather than sounds.

The V1 shape pathway goes on to the temporal visual memory area, and damage causes trouble remembering what you have seen and comparing it to what you are seeing now. This is consistent with failure of percept-to-percept-memory and failure of something like the beep-boop pathway of the auditory system.

One branch of this comparison pathway goes to the amygdala and specializes in the perception of emotional expression. This would be of use to social primates who spend time looking at faces.

What is the Brain and What is it for?

Senior researchers are prone to big-picture papers with grand titles like the one above. We will discuss such a paper, although one with a more modest title, and these two questions.

What is the brain? We have made a case for it being a large multi-modal, but mostly visual, model of the outside world.

What is the brain for? Horace Barlow, who was a major researcher and well qualified to consider such questions and write such papers, answered that it was for activities that required " ...extensive knowledge and understanding of an animal's environment". (And so, for representation of that environment.)

He asked: "Why Does the Cortex Everywhere Possess a Similar Structure?" His answer was because "it performs the same operation everywhere." (And on sensory data that has been coded in the same impulse train language.)

What is that operation? His answer was "...noting suspicious coincidences in its afferent [sensory] input and thereby gaining knowledge of the non-random, causally related, features of its environment". (Features that are useful to learn and model.)

The three center-surround cell module notes the coincidence of three activated cells and codes it as a line. The primal sketch module notes the coincidence of three activated line modules and codes it as a triangle. (It does pattern recognition.)

Barlow concluded: "The important knowledge about the environment is contained in these associations and their structure, and it is proposed that the elementary operation performed everywhere in the cortex is the extraction of this knowledge."

The extracted knowledge is stored in memory as a Craikian internal model. The next time a triangle appears, in the outer and inner world, Marr's percept-to-percept-memory module could recognize it as such and assign the input signal to the model representation. (Again, pattern recognition.)

Barlow's hypothesis fits well with Craik's. (In fairness, it should be mentioned that he wrote his paper at the Craik Research Institute at Cambridge.)

Development of the Visual Processing system

Vision is not completely hard-wired. It must learn to do its job.

In the 1980's a rather odd neural network experiment was carried out. The experimenter was Ralph Linsker, and he was interested in what the visual system did in the womb. This is a quirky question but an interesting one. The visual system is complete by the third month of gestation. It sits in the dark for the next six months with no visual input. What happens to it?

Linsker modeled vision as a multi-layered neural network: The retina was layer A with cells that were programmed to fire randomly and generate meaningless noise, as they might in a dark womb. These cells stimulated layer B, which stimulated C, and so on up to layer G. Layer A had no internal connections and could not self-organize but the other levels could.

The network was run in this state and then examined to see how it had organized itself: Layer B made center-surround cells like the output cells of the retina. Layer G made line detecting cells like those of the visual cortex.

If lateral inhibitory connections were added to the G layer, the line detecting cells organized themselves into areas responding to lines at particular orientations like real cortex. (Lateral inhibition made these networks similar to the Kohonen networks that self-organized themselves into hot-spot feature-detectors.)

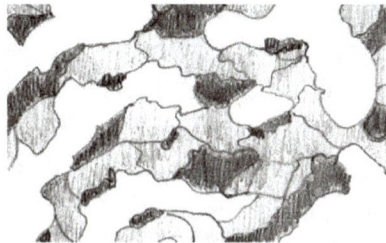

Figure 100 Layer G neural network map with dark areas indicating hemi-field columns with fine structure of darker-shaded areas that respond to lines at one angle. (After a diagram in Linsker, 1986. The hemi-fields are not in his diagram.)

A connectionist neural network comes out of a model-womb with line-detecting neurons arranged in learned patterns. A real fetal network probably does so as well but is still not ready to do its job. It must learn to see the world outside.

Hubel and Weisel did experiments on the development of real visual systems after birth:

They sutured an infant monkey's eye shut for the first three months of its life. When opened, the eye was unable to see. It had been made artificially blind.

They sutured an eye shut for only six weeks and the hemifield cortical processing stripes for that eye narrowed markedly.

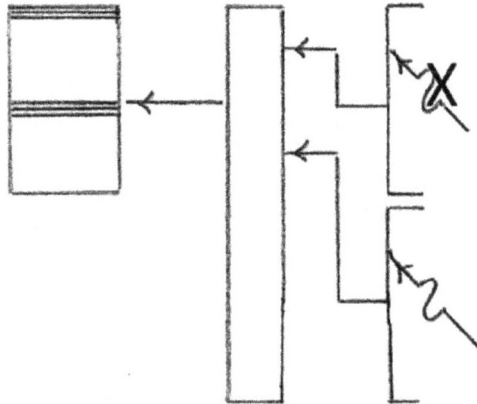

Figure 101 Thin L eye response strips in L occipital cortex after six weeks of L eye closure. (Schematic after LeVay, Hubel, and Weisel, 1980.)

Blakemore and Cooper raised infant monkeys in rooms with no horizontal lines, and the columns that registered those lines narrowed.

The visual system then has two critical early learning periods, first in darkness and then in light. It continues to learn in later life. You train the eye to see a golf ball in flight or Watteau's trees. You in fact train the brain; you train it to perceive.

Vision as a Construction

We discussed how the perception system shaped our auditory reality in the last chapter. Now we can see the processing system literally shaping our visual Umwelt.

There is no white shape below. V1 looks for edges and can manufacture illusory ones, as the phantom triangle illusion demonstrates.

Figure 102 The phantom triangle. (After Kanizsa, 1953.)

There is no animal below. A drawing can represent a great deal with a little information about edges. A few lines can be a horse—or rather, the illusion of one.

Figure 103

The power of this limited information to stimulate our visual systems suggests a link to cerebral representations. Does a pencil sketch trigger a primal sketch?

The representation system for color also constructs and distorts perception. Red can be red—but it can also be the absence of green. White does not exist at all. There are no white receptors. White is a construction of the nervous system when all three color-receptors are activated—a default signal.

Perception is also based on internal knowledge built into the representation system—the mental model. The psychologist, Richard Gregory, thought that since the visual cortex gets 80% of its input from other cortical areas and only 20% from the retina, vision should be 80% based on stored knowledge. He suggested that, if that knowledge were changed, so should the perceived object.

We can, at will, change a corner of the Necker cube from internal to external. We can, at will, see the Rubin vase or the people.

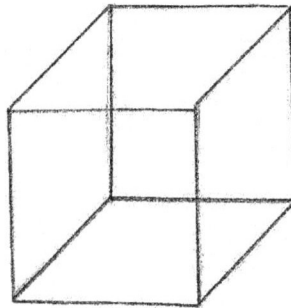

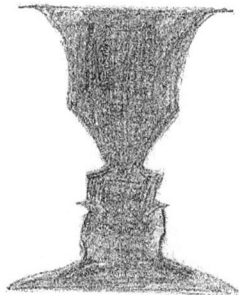

Figure 104 Necker cube above and Rubin vase below.

Gregory's Perceptual Hypothesis Hypothesis

Richard Gregory considered the oddities of visual perception and concluded that perceptions are "predictive hypotheses of the external world...[and] our most immediate reality ". In other words, perceptions are guesses about what stimulated the receptor set. Meaning is matching to the correct representation. We see meaning.

This hypothesis fits with Craik's internal model hypothesis and the phantom triangle literally illustrates it: sense data are assigned to an internal construct which is then projected out onto the outer world.

Visual recognition can be modelled as recovery of a stored content addressable memory

$$fm = M. r'$$

$$= (\Delta M' + \Delta M'' + ...etc.). r'$$

$$= \Delta M'. r'$$

$$= f'$$

and the f' in this case is Gregory's predictive hypothesis.

Complex Cortical Visual Malfunction

The multiple cortical visual maps mean there are multiple ways that things can go wrong and at varying levels of complexity. A stroke in the primary visual areas leaves a blind spot. At higher levels, there can be more selective failures.

Years ago, I drove past a tree with lavender flowers, and asked my wife what it was. She gave me a curious look and said, "Every year when the jacarandas bloom you say, 'How pretty' and ask me quite politely what they are. You've done that at least three times."

Now, when the jacarandas bloom, I do not have to ask her. Not because I remember the tree, but because I remember the conversation.

I am not alone in my difficulty. Patients with small strokes in the parietal area lose the ability to recognize fruits and vegetables but can recognize and remember everything else. I thought this was quite odd when I first heard of it, but it makes sense in an evolutionary context: fruit-and-vegetable recognition was a driving force in primate visual evolution. There would be modules devoted to it and they could be damaged or missing.

Stroke victims with damage to the parietal area can not recognize faces. One learned to recognize his wife by the clothes she wore. Again, this makes sense in an evolutionary context. Members of a social species would devote cortical map space to recognizing fellow creatures and could suffer damage it.

These are partial visual agnosias. Facial non-recognition has its own name, prosopagnosia (prosopon-face).

They can be even more complex and even more abstract. The angular gyrus sits at the junction of the areas for visual and auditory processing. If you have a stroke there, you are left with an odd set of symptoms: You cannot recognize colors, you cannot tell right from left, and you cannot do mathematics. Interestingly, Albert Einstein's angular gyrus was abnormal—abnormally large. Was his brain tuned to mathematical concepts? Is mine out of tune for flowers?

The problems of the auditory system also occur in the visual: We misinterpret what we see. We neglect one side of visual space. We see things that are not there.

People with migraine headaches have problems with the line detection system. They see phantom shimmering bars of light called scintillating scotomata. They start as small, semi-circular, jagged distortions and increase in size over twenty minutes to occupy half of the visual field, often with a blind spot (scotomata) at the center. They indicate trouble in the visual cortex of the opposite hemisphere. They fade away and as they do the headache develops on that side.

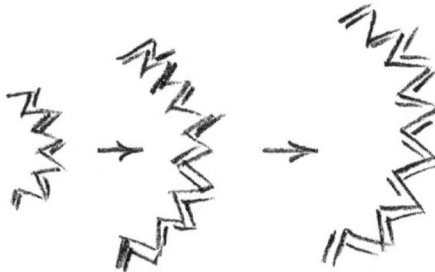

Figure 105 A migraine scintillating scotomata hallucination begins as a small, jagged scintillating pattern that grows over twenty minutes until it fills one side of visual space. It often has an area of absent vision at its center (a scotomata).

We know what they are. We can record abnormal electrical activity moving slowly across the hemi-cranial visual cortex. They are electrical disturbances experienced by the patient as hallucinations. They confirm the generative nature of the visual system: There is no stimulus. They are generated by the visual cortex.

We do not know why they happen. There is no structural abnormality. They suggest that the line recognition system can become unstable and jitter. Migraines may be the price many of us pay for high-level visual processing. Interestingly, those who suffer from them are more bothered by intense stimuli. Perhaps their sensory systems can be more easily driven into instability.

Neurologists spend a lot of time with migraine patients. Nine percent of males and eighteen percent of females have them. (The Greek, hemicrania, became the Latin, hemigrainea, and then the old French, migraine. The Greek, patior-to suffer, became the English, patient).

The Processing Perspective

Visual processing pathways can be abbreviated, like the auditory, into a primary and two further processing stages—forty processors collapsed into three.

There are connections to the auditory stages and connections to the limbic area, and to higher-level stages and one of these could be the visual percept-to-percept-memory map (M).

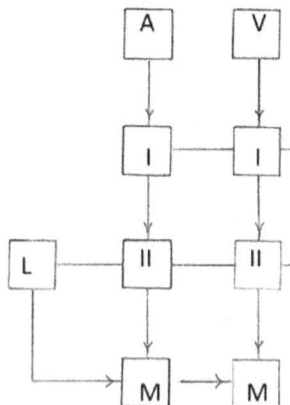

Figure 106 The parallel A and V processing stages and the percept-to-percent-memory (M) maps.

There is an area that is not shown between the two pathways that receives inputs from both A and V and is called a bimodal area. It is an abstract combination of vision and sound like the smell-taste combination representation of an earlier chapter. We will discuss this later.

Representational and Generative Perceptual Sense Organ Learning

We have visual brains: More than half of the sensory fibers entering the central nervous system carry visual signals. More than half of the cortex is devoted to visual processing. Our internal world is mostly visual.

The mammalian brain develops perceptual hypotheses about what is stimulating the receptor set and codes the sensed outer patterns into representations in a perceptual theatre. Its complex perceptual sense organs interpret inputs as relating to mental models rather than record exact pictures of reality.

This construction system is quite different from the naive concept of a projection of images through the eyes like a camera. The retinal cells send what are basically telegraph messages to the visual cortex. Forty different areas there process the messages in parallel, and somehow combine them to produce a subjective impression of a window on the outside world. That subjective impression is an illusory construct based on electrical impulse trains—"a dream modified by [visual] input" that we seem to see but in fact manufacture.

This more complex processing system allows the extraction of more meaning from the world but at the price of more complex mistakes. The generative processing system itself can determine how the world is perceived. It can generate illusory lines. It can hallucinate. It can sense but fail to understand a percept stripped of meaning.

How do we do this? How do we make sense of information from the eyes? How do we assign meaning to a pattern of neural firings? How do we make a telegraph signal denote an image in the outside world?

The Craik model solves the problem of meaning. The meaning is built into the internal representation. The only problem is the correct assignment of sensory pattern to representation. One could argue, however, that the problem is not solved but just shifted from the act of perceiving to the act of learning meaningful representations.

What can we say about learning representations? Not much, we do not know how it happens, but we can say something about timing.

In neural network experiments, the network learns representations during a training period: The network starts with a blank memory and a large learning-multiplier and is trained with a series of example signals. Once it has learned the training set, the synaptic weights are frozen (or made much less changeable by reducing the size of the learning-multiplier), and the network goes to work recognizing patterns or, to put it another way, assigning meaning to input stimuli. This initial period of learning representations is like childhood, a period of rapid learning followed by a less adaptable adult phase.

The learning continues. The cerebral representations of childhood are added to and modified throughout life. They are linked to other representations and take on deeper meanings. The internal maps increase in complexity.

The depth of the internal world affects the act of perceiving. The complexity of the map determines the resonance of the percept. Dead eighteenth-century French painters can walk with us on twenty-first-century golf courses. So, can war movies. Perhaps we should be careful about how we train the internal map.

CHAPTER 10

HOMINIDS AND HANDS AND MOTOR BEHAVIOR ORGANS AND LORENZ'S BEHAVIORAL WORLD

I can pick up a stone and use it to hammer something with my hands. I can anticipate a need and make a tool like a stone hammer with my hands. I can appreciate a threat and anticipate a need and make a hammer or spear to throw and hit something with my hands.

These abilities developed as primates evolved into hominids. Thirty million years ago, the old-world monkeys separated from our line. Ten million years ago, the gorillas separated. Five million years ago, chimpanzees separated from our hominid line.

The hominid's more complex brain could not only sense more but do more. We are going to discuss doing.

Hominids

This happened in east Africa where tectonic plate separation had led to the formation of a huge rift valley with small forests separated by broad stretches of grassy plain or savannah. This environment favored animals that could move quickly on the ground.

The primates came down from the trees and began to walk upright. That posture allowed them to see better in the tall grass and left their hands free for throwing and hunting.

The increased predator risk in the open increased the benefits of group behavior. Group hunting became possible. The hands were then free for communication by gesture.

Four million years ago, the first bipedal hominid was walking the plains of east Africa. This was Australopithecus. The most complete skeleton is known as Lucy. She had a brain weight of 400 grams.

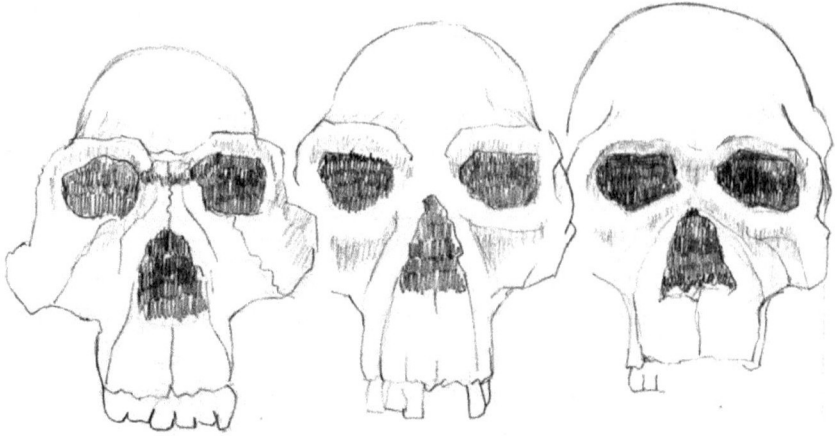

Figure 107 Australopithecus and Homo habilis and Homo erectus skulls. Note the prominent brow ridges, and the increasing size of the cranium. The chewing muscles continue to go through the cheekbone arch to the lower jaw.

Three million years ago, a new hominid, homo habilis, evolved, with a brain weight of 600 grams.

Two million years ago, homo erectus appeared, with a brain weight of 900 grams. Homo erectus was a tool-user, but his stone tools were limited to crude implements like chipped hand axes. Once developed, those tools did not change for more than a million years.

Walking and Volition and Its Problems

Walking in four-legged animals is handled by the spinal cord. There are rhythmic gait generators in both foreleg and hind leg regions with coordinating pathways running between them. A cat with its cord severed from its brain can walk if put on a moving treadmill.

When a quadruped rears up on its hind legs to become a biped, its brain becomes more of a factor in walking. Think of the volitional decisions and thoughtful control needed when walking through a minefield—or a child's room. Think of playing soccer.

The old walking centers are still there but under the control of a brainstem walking center that is in turn controlled by the frontal part of the brain. More levels of control means more ways things can go wrong, and sorting out why a patient is walking badly is one of the more difficult problems for a neurologist. Damage to the basic pathway from cortex to nerve is easy—one side is paralyzed. Sorting out problems due to the walking centers, not to mention the balance and coordination centers, is hard.

One of these is frontal gait dysfunction which is slow and "magnetic"—the feet seem to stick to the floor. Another is gait ignition failure. The patients can use their legs and walk normally once they get started. The problem is that they cannot get started. They stand with their feet rooted to the floor, swinging their arms, and swaying their hips to try to go. One patient of mine stands to attention, ports his cane, and whistles a march to get going. This works every time (he used to be a marine), but without it, his feet are glued to the floor.

This is a failure of voluntary motor control, a failure of volition. Volition is as tough a subject as there is in neurology. Gait ignition failure has fascinated me ever since medical school. I thought it would lead to a breakthrough in understanding volitional brain function, but it has not.

It is complicated. It does not occur with focal damage but with diffuse dysfunction, such as Parkinson's disease, which is essentially a slowing of the all of the motor brain. Also, it involves areas outside the movement system. When such patients have finally been able to start walking, a visual difficulty like trying to follow a pattern on the floor or go through a narrow door, can stop them in their tracks. There is dysfunction of the system for visual control of walking.

Volitional movement is not just complicated. There have been unusual research findings and bizarre conclusions. Benjamin Libet in 1985 measured an evoked potential related to movement and found that it started before the subject had decided to move. This led to a dispute about whether volitional movement can occur. Some concluded that it cannot, and that free will is an illusion. The argument is still going on.

I will point out that Samuel Johnson would have kicked a rock and said, "I refute it thus". This how he responded to Bishop Berkley's contention that external reality was an illusion and only mental events existed.

Hand Control and Its Problems

Hand movements are more complex than foot movements and harder to evaluate. The patient is asked to move his hand in a straightforward way, like turning it over. Then, in a complicated way, like waving goodbye. Then, in a more complicated way, like using a tool such as a hammer. If the patient played the piano, even higher levels of movement composition could be tested.

Patients with damage to the motor centers cannot do these things. They can use their arms and hands, but only if they "do not have to think about it".

These disorders are like the complex perceptual failures, where sensations register but patterns do not, in that simple movements can be made but complex ones cannot. A metaphoric explanation is that movements are symphonies—orchestrated compositions of many muscle actions—and this is damage to the score.

They are called apraxias (praxis-action). There are different levels. Holding a cigarette is a simple action, but playing the piano is complex enough to be considered motor thinking.

Gait ignition failure is also called gait apraxia, although clearly it is more than just an apraxia.

Movement and Touch and the Movement Behavior Model

The final stage of motor control is an upper neuron in the cortical motor strip telling a lower motor neuron to tell a muscle cell to contract. Before and during movement, however, motor planning and motor control systems must act, and act in accordance with sensation.

Sensation plays a major part in movement. Movement must be sensed. The world in which the movement takes place must be sensed. Movement and touch are close companions in the nervous system.

The motor and touch representations for one half of the body are side by side in the brain, with touch representations in the parietal sensory strip and motor representations in the frontal motor strip. The strips send axons to one another across the central sulcus. They talk to one another.

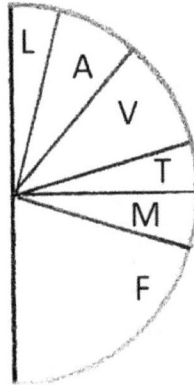

Figure 108 The frontal lobe (F) with its motor strip (M) beside the touch sensation strip (T).

A motor action must be planned to fit the sensory pattern to even grasp a pencil. The perception-matching model can be combined with a motor behavior selection model that matches the percept to motor memories and selects one to activate the motor system to hold a pencil or hit a nail with a hammer.

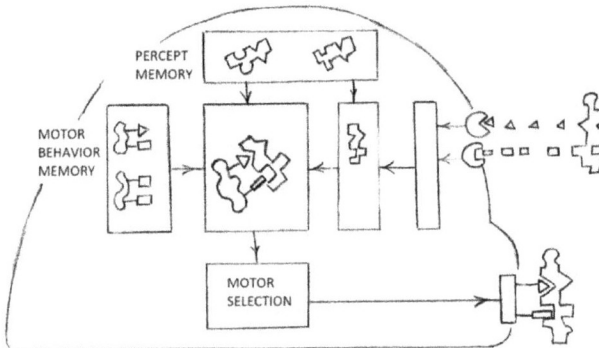

Figure 109 The Motor Behavior Selection Model. A sensory percept goes to the percept-motor-matching-module to be matched to a motor memory representation.

This system could fail. A damaged motor memory could cause apraxia.

If we model the motor system as part of the brain activity vector that can learn and remember its motor response (bm') to input (r'), so that

$$bm = M.\ r'$$

$$= (\Delta M' + \Delta M'' + ...etc.).\ r'$$

$$= bm'$$

then a matching failure could evoke the wrong response (bm'') or no response. The model could also demonstrate apraxia.

The engineering graduate student often makes a mathematical model of a real-world system. The student's prayer is: May my system be linear and time invariant. Time-invariance means that the matrices stay the same and so always produce the same output to the same input. The model brain does not do that. The model matrices are time-varying in both sensory input and motor output stages.

The model is not only not time-invariant but also only a linear approximation to a non-linear brain. The graduate student's prayer is doubly not answered. Since this is an engineering student, it is still possible to pray for linearity to a first approximation—but not time-invariance if it is going to learn.

Movement and Touch

Even after being planned and initiated, ongoing movement control is necessary. It can be controlled by vision or hearing but is particularly controlled by and reliant upon touch.

Imagine that you are in a room with no light and no sound, and your touch nerves have been anesthetized. Could you move? Yes, you could, your motor nerves would still work. Could you move and accomplish anything? Probably not.

Movement is dependent on touch sensation maps in the sensory strip. The touch representation of the body surface is a distorted picture of the opposite half of the human body. The map is distorted because areas with a lot of receptors are bigger; the thumb and lips are huge. If the thumb area is given an electric shock, the shock is felt not in the brain but on the thumb. This was how Wilder Penfield, a neurosurgeon, found and explored the sensory map. He shocked the exposed sensory areas of awake patients. Penfield called the map, the sensory homunculus or little man.

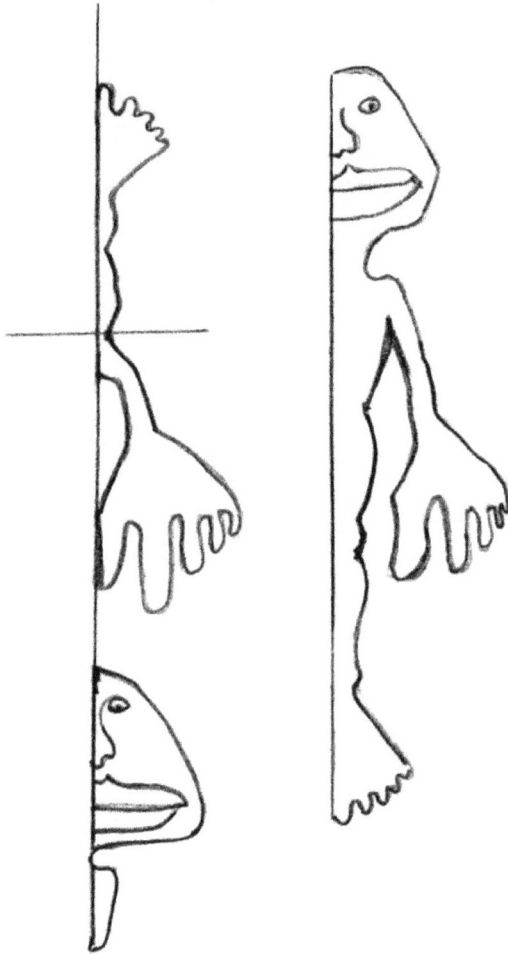

Figure 110 The sensory homunculus to the right is a distorted representation of the contralateral half of the human body. To the left it is shown as it is on the cortex with the arm and trunk upside down and face and tongue below, and the leg hanging over the top and down the inside of the hemisphere. (After Penfield, 1950.)

How does it develop? The wiring to the sensory map is fixed, but a neural network experiment by Helge Ritter in 1990 suggested that the representation is also self-organized. In the experiment, a neural network shaped like a hand sent thousands of random finger signals to an unstructured and unsupervised cortical network. The cortical network became a map with individual finger representations. This suggests that the real homunculus is, at least partially, organized by incoming sensory signals. It also suggests that it can change.

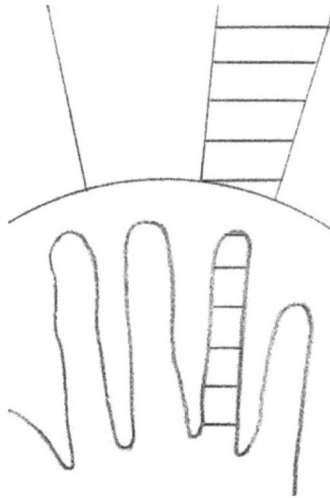

Figure 111 Schematic of the Ritter experiment with a self-organizing neural network map in the upper part of the drawing showing simulated cortex for four fingers after 10,000 stimuli with 800 hand stimulus points connected to each of 16,384 points of the cortex neural network and with the fourth finger representation cross-hatched. (After Ritter, 1990.)

The homunculus is a representation of part of the world that touches us. It is also a representation of the part we touch it with. What happens if you change that?

Although most work on perception and memory is highly abstract, an experiment done by Michael Merzenich was shatteringly direct: Take a monkey and map the cortical representation of the paw. Now cut off one digit!

What happens to the brain representation? The answer is that the adjacent fingers take over most of the now disconnected cortical areas; the map in the brain changes to reflect the new body.

Ritter reproduced the Merzenich experiment results with her network.

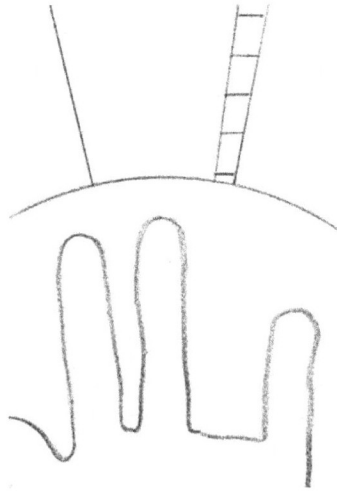

Figure 112 Schematic of Ritter's simulation of the Merzenich experiment with the fourth finger computationally amputated, and its cortical space reduced and adjacent finger spaces expanded. (After Ritter, 1990.)

The blind are thought to have particularly acute hearing. It seems more likely that their hearing is no better than anyone else's, but that their processing space has expanded, and they can bring more cortical computing power to bear.

Touch and Touch Neglect and Complex Touch

There are five primary touch receptors registering simple forms of touch, including light touch and joint position sense.

Light touch feels the outer world. It is registered by pressure sensors, most of them concentrated in the fingers. It enters the spinal cord, relays to a second neuron, crosses to the other side, and travels up to the thalamus where it relays again and goes to the sensory cortex.

Joint position sense feels the inner body. It detects joint and muscle tensions and so how the limbs are oriented. It follows a different pathway up the same side of the cord, relays in the brainstem, then crosses to the thalamic relay, and on to cortex.

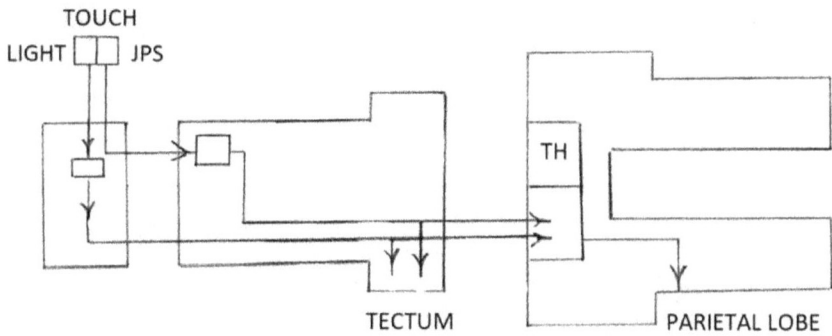

Figure 113 Two Touch Pathways: Light touch relays in the spinal cord, crosses to the opposite side, ascends to thalamus, and goes to parietal sensory cortex. Joint position sense (JPS) ascends without crossing to relay in the lower brainstem and then crosses to join the other pathway. Branches from each go to the tectum.

Damage to touch cortex can cause tactile sensory neglect. After a stroke, a patient can completely ignore touch on one side.

Both world and self are represented together in the sensory strip and tactile neglect can be odd. Not only can the representation of the outer world be lost, but the representation of the body as well. The brain can ignore that side of the body. It can treat it as a foreign object. If the neglected arm is brought across and placed on the good arm, the patient can think it belongs to someone else. Perception of the body can change to reflect the changed brain map.

Beyond simple touch, there are complex synthesized touch perceptions in the cortex. Some require movement: If you place your fingers on wood or paper or leather, they feel the same. If you run your fingers over them, they are clearly different. Once again, the motor cortex talks to the touch cortex.

Complex touch is pattern recognition. If the touch cortex is damaged, simple touch remains intact but complex pattern recognition is lost. Patients can no longer identify textures or objects like coins or numbers written on the skin. This is tactile agnosia.

Joint Position Sensation and Movement Representations

The messages about muscle and tendon stretch let us know where our legs are, which fingers we are using to scratch, and how to type without looking. If these were anesthetized, a person in a dark room would not know where his limbs were or whether they were doing what he wanted. He would have no touch control of movement.

They do more. One dark night, I and my fully functioning nervous system went out to the balcony, found one of the black chairs by touch, sat down—and fell!

I dropped—and stopped—four inches down. Someone had taken the cushion off the chair.

My internal world model contains models of my chairs; and, for each one, a prediction of the exact joint position at which I can relax my leg muscles and hit the seat.

Every movement I make is dependent on these models. If they go wrong at any level, my motor control system cannot do its job properly.

Motor control, like sensory processing, is a hierarchy of processing stages, but each depends on sensory input and fails if it is wrong. If the leg peripheral nerves are damaged and the low-level joint position information is wrong, then I walk awkwardly and fall when I close my eyes and lose visual guidance. If the high-level, sensory-motor memory no longer matches the outside world, I land with a thump. The movement of the body through the world and the world it moves through are woven together in the brain.

Sensing and Action and Its Representations

Imagine that you wake up in total darkness. You reach up and feel a wooden surface close above. You reach out to the sides and feel wooden walls that seem to taper down toward your feet. You touch your chest and feel a large flower: A lily. You are buried alive! You are stuck in an Edgar Allen Poe story!

Before you started to hyperventilate you were able to use arm position sensations to map the oblong box around. You were able to use the internal representation of your body to create a three-dimensional representation of your immediate world. Perception here was as much a motor act as a sensory one. Again, the motor cortex talks to the sensory cortex. We think of sensation and action as distinct and different, but they are woven together in the brain.

Brain activity imaging shows differences in the representations of man-made and natural sensory objects: Natural objects are found in the parietal sensation areas as expected, but man-made ones are often in the frontal regions.

The frontal lobes are concerned with doing. Man-made objects are often concerned with doing. Their names reflect this by acting both as nouns and verbs: hammer, saw, knife. The brain treats them as more action-like than sensation-like. Words and actions are woven together in the frontal lobes.

Later when we start to map the world with such words, we will think of them as sensation-like, which is to say we will put them into conceptual categories the brain does not fully agree with. This can lead to problems: Where does your fist go when you open your hand?

The Processing Perspective

The neocortex is now a multi-modal sensory display and multi-part action area.

Pandya and Seltzer's touch pathway parallels the other pathways with inter-connections at higher levels. One of those connections would be to a touch percept-to-percept-memory-matching module (M).

Another higher-level connection is to a multi-modal area where three sensory representations are combined into a trimodal representation, a touch-sight-sound. What the brain does with such an abstract representation, we will discuss later.

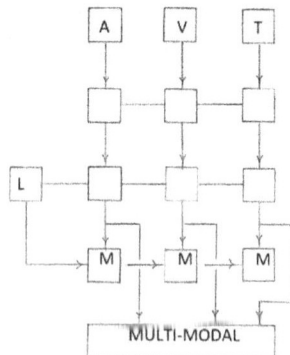

Figure 114 The three sensory processing pathways and their connections. The M-area on the right is the touch-to-touch memory area. The multi-modal area is at the bottom. (The bimodal areas between the three pathways and their connections to the multi-modal area have been left out to make the diagram readable.)

Pandya and Selter note that the cortical motor areas parallel the sensory processing areas with connections at higher levels. These levels include the module for percept-to-motor matching which is shown in the next diagram as a separate (M) level in frontal cortex. (The M and multi-modal boxes are not in the Pandya and Seltzer paper.)

The fastest inputs to the motor areas are touch and joint position. Their sensory strip lies side by side with the motor strip, hand by hand and leg by leg. Their connections are the shortest. Both use egocentric maps. Touch and movement work closely together.

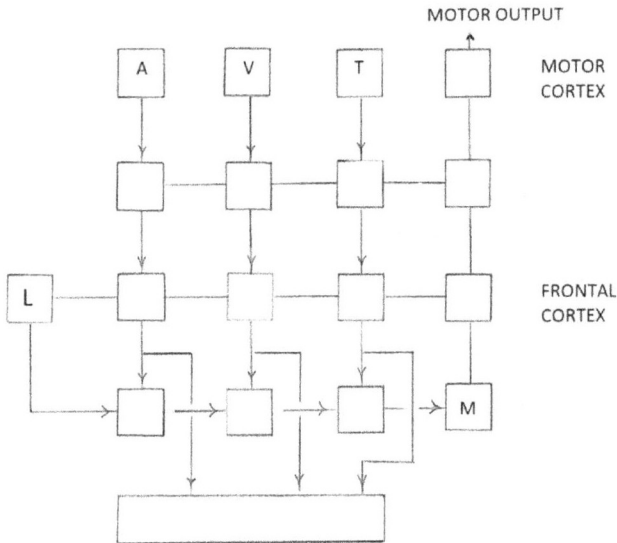

Figure 115 The parallel sensory and motor pathways with the motor planning pathways flowing from bottom to top with motor cortex exiting to motor output areas in brainstem and spinal cord . The frontal motor areas are higher-level motor planning stages and include the percept-to-motor matching module (M) .

The diagram above is complicated. It is a combined representation of the Pandya-Selzter concept, the Marr percept-matching concept, the percept-to-motor matching concept, and the multi-modal representation. Even so, it is an incomplete representation of motor control and its sensory inputs. It omits the hearing and vision-directed movement systems, the spinal cord reflex system, the brainstem reticular movement system, the brainstem system for postural control, the cerebellar system for coordination, and the diencephalic system for motor activation and planning that is damaged in Parkinson's disease. There is much more to motor control, but it is not the subject of this book.

Lorenz's Motor Behavior World

Sensing the world and moving the body through the world meet in maps inside our heads.

These maps are egocentric. In them, the boundaries between in here and out there blur. Sensation and action blend. The words blend: Hammer means object and action.

Our sensations frame and direct our actions. Our actions change our sensations: Acting in the world, even just standing up, changes our perception of it.

We exercise volition and choose motor behavior tools. They change how we act upon the world. Konrad Lorenz pointed out that a motor behavior is like a paw or a beak: It allows action directed at the outside world. A behavior is a motor organ.

Such volitional motor behaviors are not pre-programmed but made. We train ourselves to walk or use a hammer. We train ourselves to throw a ball or to dance. These are new things in the world. Baseball and ballet did not exist until we created them. A behavior is a learned motor organ.

We manufacture motor organs and change the world we inhabit with them. This Lorenzian world is one of our creation.

PART IV

HOMO SAPIENS AND CONCEPTUAL MODELS

CHAPTER 11

HOMO SAPIENS AND DIRECTED ATTENTION AND CONSCIOUSNESS AND POPPER'S PRE-SELECTION WORLD

Around 500,000 years ago, the first homo sapiens appeared. They were not us; they were the old model, homo sapiens archaic. They came in different versions with different brain weights and included the Neanderthals with brain weights of 1450 grams, which is more than ours.

Around 300,000 years ago we appeared: the new model, homo sapiens sapiens, with an average brain weight of 1350 grams.

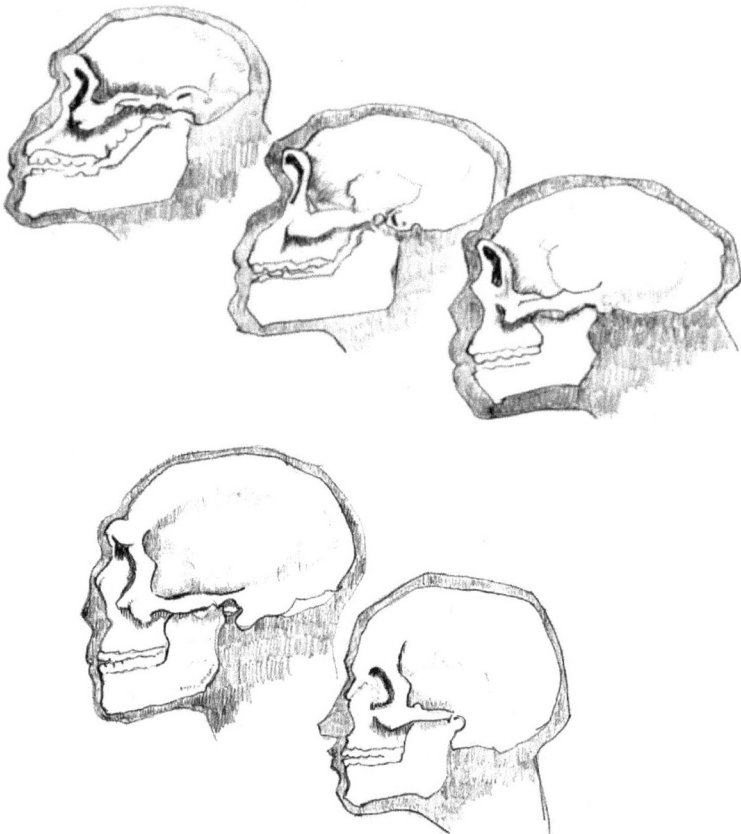

Figure 116 Australopithecus, Homo habilis, and Homo erectus on top. Homo archaic neanderthalensis and Homo sapiens sapiens on the bottom. The enlarging brain changes the head, and the enlarging frontal lobes change the forehead slope.

What is different about the homo sapiens sapiens brain?

It is relatively bigger. It has more available to do things other than run the body. Its ratio of dispositional to total brain weight is the largest, as Aristotle was the first to point out. The size difference is not apparent at birth. The infant human has a 350-gram brain like a chimp. The chimp's period of maturation is short, and the brain grows only a hundred grams. The human period of childhood and brain growth lasts more than a decade, and the brain weight quadruples. The price of that growth is prolonged childhood vulnerability; it can only happen if mother sticks around.

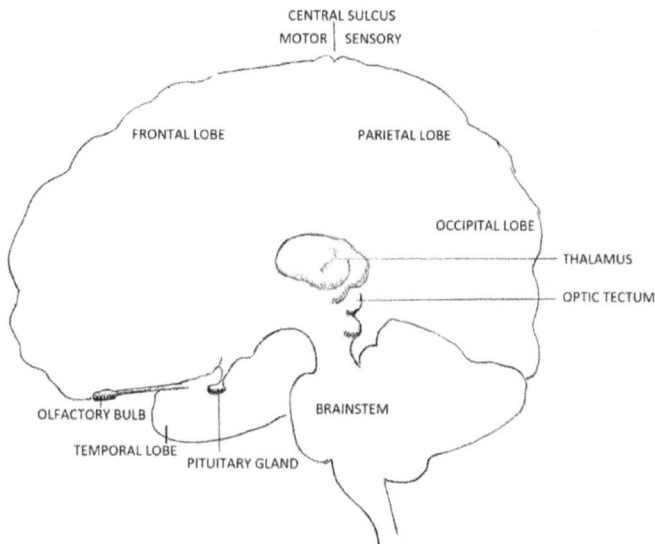

CENTRAL SULCUS
MOTOR | SENSORY

FRONTAL LOBE   PARIETAL LOBE

OCCIPITAL LOBE

THALAMUS

OPTIC TECTUM

OLFACTORY BULB   BRAINSTEM

TEMPORAL LOBE
PITUITARY GLAND

Figure 117 Human 1350-gram brain in mid-section. The frontal lobe is forward of the central sulcus. The brainstem optic tectum is just below its thalamic connection which is the bump at the back of the thalamus (lateral geniculate body} that projects to the occipital visual cortex. The olfactory bulb is relatively tiny, and its olfactory cortex is in the inner aspect of the temporal lobe.

It looks different. It has larger neocortical areas, and they change the shape of the head. It has larger frontal lobes, and they change the shape of the forehead.

It works differently. The human frontal lobe is relatively bigger than that of the primate, but much of the size increase is in myelin cells and so in processing speed. The progressive myelination of the frontal lobes parallels intellectual and emotional growth through young adulthood into the thirties.

The human brain is a bigger hammer. The human frontal lobe is a bigger and faster one. It is, of course, more than just that.

Frontal Lobes and Delays and Working Memory

How does it work? What are the frontal lobes for?

A frontal lobe damaged chimpanzee has trouble with delay tasks: If it watches while a banana is hidden, it has no problem finding it. If it is made to wait, it cannot find the banana.

A frontal lobe damaged human has trouble with a test called the Wisconsin card sort: The patient is shown cards, one at a time, and tries to guess how to sort them. With each card sorted comes a yes-no response. After realizing that the sorting is by color, the rules change, and the patient then must work out that the sorting is by number. The purpose of the test is to see how rapidly the patient can shift strategy or mental set, and the frontal lobe patient does not do well.

A frontal lobe damaged human can change personality and become impulsive, and thoughtless. The famous example is Phineas Gage, who had a railway tamping rod driven through his frontal lobes in 1868 and became "… not the same Gage". He changed from a sober, inhibited Victorian into a blasphemous, impulse-ridden, free spirit: sort of a Victorian hippy—a child of nature, but not much fun to be around.

This is not the only frontal lobe syndrome, but it is the most interesting. The most common type of frontal lobe patient is dull and apathetic. This is usually what happens after a frontal lobotomy.

The frontal lobes deal with delays and shifting mental models. They compute responses that are not stimulus-driven, that cannot be handled by rote, and that require consideration before action.

They get information from other areas of cortex. They hold object representations in a frontal short-term system called working memory. They delay action as they engage in complex processing of contingencies and selection of responses. They consider the state of the inner representation of the world around them. They are the probable location of Craik's internal model.

Frontal lobe function is called executive function. Working memory, delays, and selecting strategies are key components, but equally important is stimulus-response suppression. Frontal lobes must inhibit stereotyped responses to allow complex processing time to run to completion. When damaged, they cannot prevent stimulus-driven behavior. Frontal lobes inhibit reflex bladder emptying. Frontal lobe patients just let go. They are incontinent—of urine in this case, and of emotion in Phineas Gage's.

Frontal Lobe Electrical Measurements and Memory Organization

The P300 evoked response is linked to frontal lobe function. The system stores an expected or routine tone in memory, and signals if an input tone is not the expected one. We called it a mismatch indicator earlier, but it could also be thought of as a novelty detector. A missing novelty detector might make the world seem somewhat flat and uninteresting—an endless succession of similar tones. This could contribute to the apathy and indifference of frontal lobe patients.

Novelty can only be appreciated by a brain with a well-organized memory. The olfactory bulb model stores all stimulus memories and blends them into one. It explains how the thousand times you have driven to work can be collapsed into a general memory with yesterday's drive barely remembered, but not the vivid recall of the time you had the car accident. The content-addressable model could remember that novel accident, but only if you were in the same place—unless it used non-egocentric coding.

Neither model can begin to deal with the complexity of human memory storage. Human memories are highly organized. They form a usable library, not just a room filled with books. They are indexed and catalogued and linked to other memories.

Novelty is one way of indexing and organizing.

Multi-modal Sensory Areas and Synesthesia

In the bimodal areas of the primate and human brain more than one sensory modality causes activity. In the trimodal areas, all three do. There are three multi-modal areas: one in the sensory cortex of the auditory lobe, one in the frontal lobe, and one in the limbic lobe. The brainstem tectum is an older multi-modal area.

They connect. In particular, the sensory area projects to the frontal.

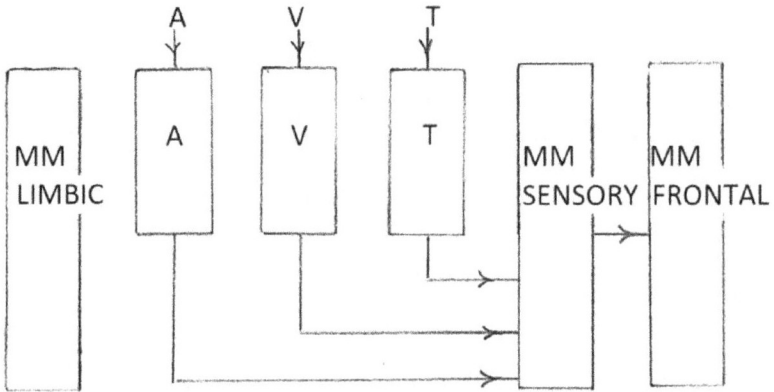

Figure 118 Sensory processing streams are collapsed into single boxes connecting to the multi-modal sensory area which projects to the frontal multi-modal area.(The bimodal areas are omitted. The limbic connections will be shown later.)

What are multi-modal percepts like? We do not know. We do not experience them. They are not touch or sight or sound. They are more removed and more abstract, more like descriptions than perceptions.

We do not know what they are like, but we have manufactured one, or at least a bimodal one: When we were talking about fish, we combined a smell-vector with a single digit from the taste system to make a multi-modal representation.

The vector representation makes it easy to do this. It also makes it easy for a single modality input to trigger the reconstruction of a complex multi-modal representation. Even a partial or degraded input could do it.

Is this what the multi-modal-representation is for? Is it the unexperienced but deciding percept? Could it project back down to the primary areas to bring them all into synchrony? Top-down directing bottom up? Is perception like a set of reflecting mirrors?

What about multi-modal pathology? Again, we do not know. Neurology has not identified a syndrome of being unable to make multi-modal representations. Multi-modal agnosia is not a diagnosis.

There is a clinical syndrome that may be relevant. It is called synesthesia (syn-together, esthesia-sense): One sense evokes another. A touch has a smell. A sound has a color. The philosopher, John Locke, in 1690, wrote of a man who saw red when he heard a trumpet. In a more complicated form, a number has a color.

Is this multi-modal perception? Maybe. A sound evoking a color is clearly multi-modal, but there is no blended percept.

Is multi-modal area dysfunction the cause? Maybe. But it could just be contamination of one sensory area by another.

These are speculations, but synesthesia lends itself to speculation. It is an arbitrary linkage of unrelated sensations. It is an abstract relationship created by the nervous system. It is odd. It is like a metaphor or an analogy. It has been speculated that it could relate to creative thinking.

It has been speculated that it could relate to the origin of language. Multi-modal representations are abstractions like words. Did synesthetic cross-connections contribute to word development?

Figure 119 Kiki? Bouba? (After Ramachandran and Hubbard, 2003.)

Ramachandran and Hubbard asked experimental subjects to match the two objects shown with the nonsense words, "kiki" and "bouba". Ninety-eight percent matched "kiki" to the angular object. It just felt right. They wondered if this sort of "feel" underlies naming? They speculated that we may all be closet synesthetes.

A final speculation: Could this be sensory processing in evolution? Is the brain in the process of creating new ways of interpreting the world? Perhaps associating a number with a color will be of use for technological man and we will select for it. Perhaps we are evolving into synesthetic perception.

Directed Attention and Common Sense

Do our brains do anything with these unexperienced representations?

Aristotle anticipated this question. He thought that the various senses were combined into a single sense, which he called the common sense. He thought that only the common sense went on to act on the rest of the nervous system.

He was mostly wrong. Sensation is processed in multiple pathways and stages, and each can project to and act on other areas.

He was partially right. We do have a common sense and it does act on the rest of the nervous system: It directs attention.

You can test direction-of-attention simply by attending to the feelings of your left big toe. Although you were unconscious of it before you read this line, you can now appreciate that the sock is applying pressure, mostly to the front of the toe, and that the pressure increases if you move it. How many times today were there pressure changes that never made it to awareness?

You have just directed attention to a touch pathway. You can do the same to other sensory pathways as well: You can tune into a conversation at a party. You can focus on this line of type.

James Skinner and Charles Yingling in 1977 found that directed attention depended on an area at the front of the thalamus. Sensory signals go through the thalamus on their way to the cortex. It is the logical place to control sensory input.

The controlling pathways start in the sensory multi-modal area and project to the frontal multi-modal area which projects to a multi-modal area at the front of the thalamus called the nucleus reticularis (net-like). Its neurons project into the thalamic relay nuclei and control sensory throughput to the cortex. You just used it to control toe-sensation.

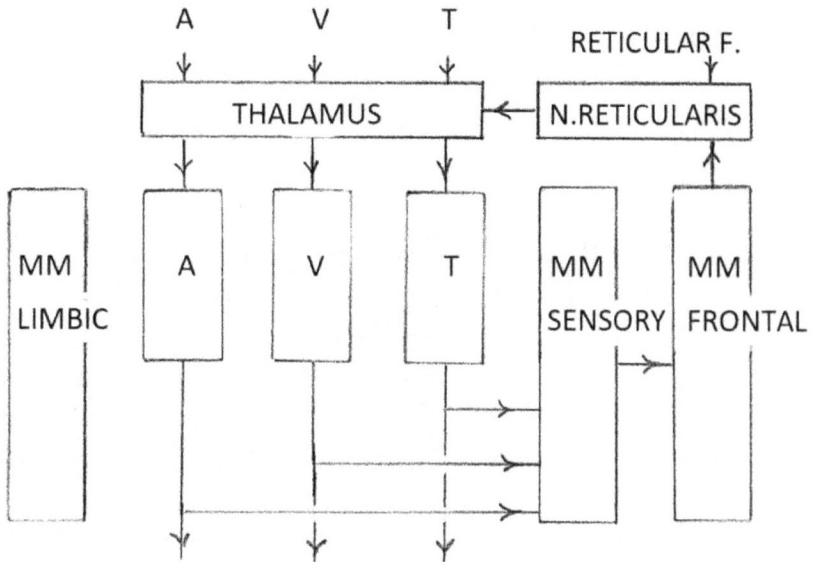

Figure 120 The nucleus reticularis at the front of the thalamus controls sensory information passing through the thalamus. (After Skinner and Yingling, 1977.)

The frontal lobe is not the only control area. The old reticular formation of the brainstem can also affect sensory throughput. As you would expect, it does so more crudely. When it takes over the nucleus reticularis, all sensory modalities are sent to the cortex at high intensity: The system is on! Think of the vividness of sensation when you are startled or frightened. The frontal lobe does fine control. The reticular formation, gross control.

A candidate for a disorder of directed attention is attention deficit disorder (ADD), which can be thought of as inability to suppress responses to distracting stimuli and concentrate on the task at hand. It is a childhood developmental disorder associated with changes in many areas, including the frontal areas, so it is not a pure disorder of the directed attention system although that is likely to be involved. The condition affects five percent of adults, and a larger percentage of children.

Measuring the Attention System

The attention system can be recorded in action. Earlier we discussed the evoked responses to auditory tone stimulation of the brain. We were interested in the P300 but there was an earlier wave at one hundred milliseconds called the N100.

The listener can volitionally change the size of the N100. Steven Hilliard told his subjects to pay attention only to tones in one ear, and the N100 waveform was larger on the side of the attended tones. The N100 reflects the act of attending.

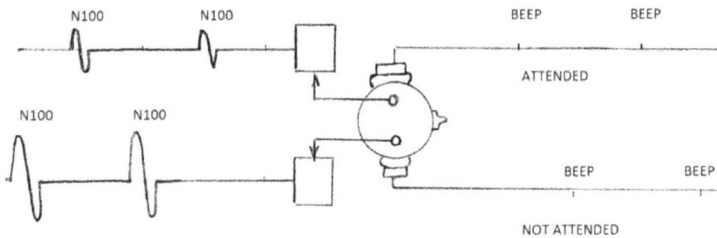

Figure 121 Attention and the N100: The subject listens to beep sequences through earphones to each ear but attends only to those on the left going to the right side of the brain and those N100 responses are larger. (After Hillyard, 1973.)

The N100 is affected by damage to the parietal and frontal areas that connect to the nucleus reticularis, which suggests that the directed attention system may have to do with it. It would be interesting to know what happens after a small stroke in the nucleus reticularis. I once had a patient with just that stroke, but she refused the test. As far as I know, no one has ever done the experiment.

Winner-Take-All Switching Brain Model

We can incorporate multi-modal sensory stages into our brain model by feeding primary perceptual processors into another neural network stage. We do not have to worry about designing multi-modal representations; the neural network will create them by itself.

We could make the model synesthetic by back connecting the multi-modal stage to each primary sensory stage. We could also do it by cross-connecting the primary stages. There does not seem to be any clear use for this, so we will not model it.

Directed attention has obvious uses. We could model it as a multi-modal processor that, rather than integrating sensations, simply compares them and designates the most active sensory stream and directs the nucleus reticularis to open that gate. The sensory stream could in turn direct brain output to a motor program and so control behavior. Moving prey could activate a visual-to-head movement stream, and so on.

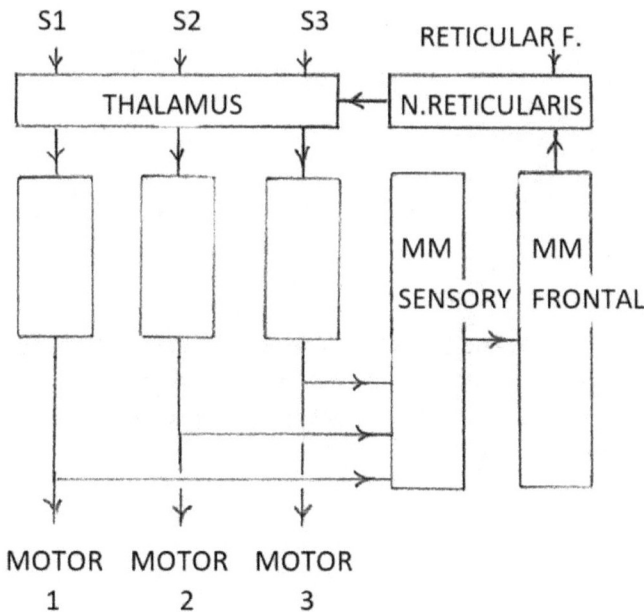

Figure 122 Competitive-attention Model: The multi-modal areas select the most active sensory stream and direct the sensory gating system to allow it to activate its motor output.

Attention deficit disorder (ADD) often involves hyperactivity and impulsive behavior and then goes by the acronym, ADHD. This could be thought of as an inability to suppress motor responses to stimuli, and the model suggests how the two sets of symptoms could relate.

The model is a competitive selective-control system, a winner-take-all system. The most exciting sensory input wins and perpetuates itself as the dominant sensory input until something else becomes more exciting and control moves on. It would solve the switching problem of the earlier chapter.

It is a conditional switching mechanism. If the condition of one stream being stronger is satisfied, it sets its internal state which sets the nucleus reticularis.

It is an internal state system like the electrical engineering model in the chapter on hearing. The inputs and the internal state determine the output.

It is a finite-state automaton in computing theory. It makes automatic decisions determined by a finite set of internal states, as a subway turnstile opens to your hand only if you have put in a coin and set its state to "decide" to let you do it.

How much of brain activity could be explained by a system like this?

Winner-take-all plus Limbic Motivation Model

We could add another level for limbic motivations. If the animal were hungry, it would set the amygdala selector for food. If the animal were feeling another appetite, it would set it for finding a mate.

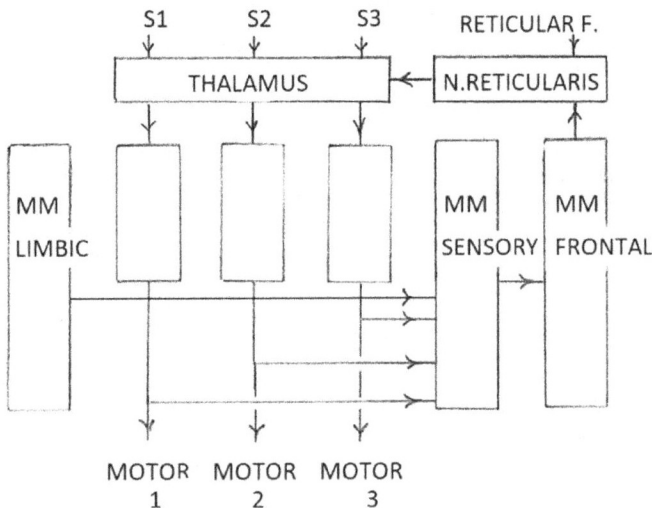

Figure 123 Competitive-attention plus Limbic Motivation Model

Adding yet another level would allow yet more complex decisions: A superior limbic automaton might register threats. If it were activated by a predator signal, it would switch on a competing fight or flight decision automaton. If not, it would allow the hunger versus mating automaton to act.

How much of brain function could be explained by layered automatons?

Functional Homunculus Brain Model

What could be done by a hierarchy of multiple layers? This idea is associated with the philosopher, John Haugland. His processors have had various names but are now called functional homunculi. They are essentially finite state automatons at different levels of decision complexity. Upper-level homunculi would have tasks that they would sub-divide into simpler tasks which they would assign to lower-level homunculi. These would in turn sub-divide their tasks and assign them to even lower-level homunculi. At each level, the homunculi would compete for control of the level lower down.

Daniel Dennett called this brain function by committee and suggested that it is all we need to explain the workings of the human brain. This may or may not be the case, but we can certainly ask: How much of brain function could be explained by committees of functional homunculi? Could consciousness?

Consciousness and Coma and Persistent Vegetative State

What about consciousness? What about the structure that allows attention and volition and houses the mind?

It is a mystery, but we can say a few things about it.

The first and simplest component of consciousness is wakefulness, the result of activation of the brain by the reticular activating system. Every creature with a reticular activating system is awake and conscious in this sense, and every such creature can be devastated by damage to it. Damage causes profound loss of consciousness or coma. The brain may be in perfect order but if this grape-sized area of the upper brainstem dies, the brain never responds to the outer world again.

The other way to induce coma is to destroy both halves of the brain. This is the reverse of the first type: The brainstem is intact and trying to activate a brain that cannot be.

A second component of consciousness is awareness. The attention-directing system is an awareness system. It allows perceptual focusing by internal control of brain activity. Control implies a controller, and control of awareness of sensation is a major aspect of consciousness.

A related aspect is awareness of stored representations or memories. The brain can contain a representation of a telephone number but be unaware of it until the number is recalled and the representation enters awareness. A system like the perceptual attention-directing system could also work as an internal attention-directing system. Francis Crick suggested this and likened it to a "searchlight of consciousness" that illuminated patches of brain maps. Consciousness involves controlling the searchlight and flipping the focus of attention back and forth between the outer and inner worlds.

Crick and Christof Koch suggested an area called the claustrum as the seat of conscious awareness. This is a multi-modal strip of neurons in the temporal cortex whose function has long been a puzzle. Their hypothesis was that it could blend sensory inputs together to create a unified percept like Aristotle's common sense and be the seat of consciousness. The most compelling evidence was found later when a patient had a stimulating electrode placed in her left claustrum: She could be flipped back and forth from normal consciousness to unresponsive wakefulness at the flick of a switch.

Only one such case has ever been reported. More common are patients who awake from coma but not to awareness. They have sleeping and waking states, and when awake gaze about—but do nothing else. This is called persistent vegetative state and is the partially recovered state of the second type of coma with bilateral brain damage and a preserved reticular activating system. The cortical damage is usually extensive, and the claustrum is not always or particularly damaged. The most noticeable feature of these patients is their lack of a searchlight of sensation or of memory: They gaze about without ever stopping to attend. They do not seem to do any internal switching. They do not seem to reflect or think or have any interior mental function. They do not seem to have conscious minds.

Why do the rest of us have conscious minds?

Do we not have to? There must be one most important thing going on in the brain at any one time. Attending to the most important sensory stream is something like conscious awareness. The attention system could evolve a similar mechanism for attending to memories. It could evolve into our form of conscious awareness.

Why do we have unconscious minds?

Do we not have to? We cannot attend to everything at once. The brain is a parallel processor with multiple sensory streams all going at the same time; most must be ignored.

Why can we not access all that is going on in our brains?

Do we not have to not be able to? Consciousness is a high-level processor working at a high level of abstract representation, so the code and content of lower-level processors would be unreadable. Also access to some areas could be counterproductive: Access to your running program might slow it down too much. You might end up like the centipede in the ditch, trying to remember which came after which, and forgetting how to run.

Conscious and unconscious minds may be not mysterious entities whose existence needs to be explained, but simply necessary parts of a brain of a certain complexity.

The Processing Perspective and the Brain Model

The sensory processing streams converge into multi-modal representations. They control sensory input, and this allows them to exert control over brain outputs including motor outputs.

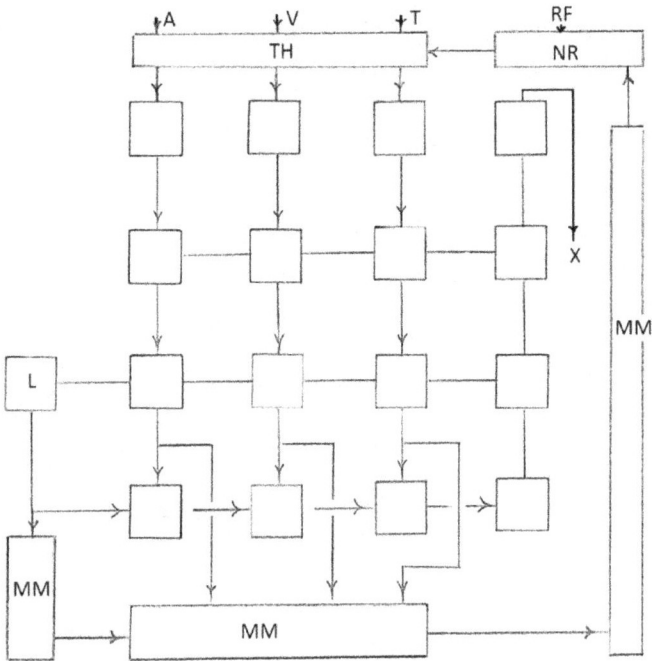

Figure 124 The model brain with sensory processing and sensory control systems and with the motor system and motor output (X) to the right.

There are a number of other control systems, including the conscious volitional system. These are not shown in the diagram, but even so it is the most complicated model of the brain in this book.

This really is a model of the brain. Even the circle diagram with sensation entering above and motor output leaving below was a model of the brain, just not a very good one—and neither is this. It is somewhat true and somewhat a lie, and the lie needs to be qualified.

It is a superficial model. It is a block diagram and more a way of organizing a model than a model itself.

It is a speculative model. Although plausible, it is not the established model of brain function.

It is an incomplete model. Most of the motor systems are absent. Only part of one of the four-part A-B-C-D brain network is shown. Many areas of the sensory processing systems are missing. In particular, the multi-modal system is incomplete; the bimodal areas and connections are not even shown.

The multi-modal representation system is of major importance. Somewhere in your brain is a multi-modal representation of your mother: How she looks. How she sounds. How she feels. How she smells. Sensory systems function in the womb, and it is conceivable that the representation includes vestiges of how she sounded and smelled from the inside. (A number of sweaty-T-shirt studies have shown that smell, even if we are unconscious of it, affects who we find attractive and even who we marry; and mother was your first olfactory input.)

This multi-modal representation is the inner model of your mother. It occupies a large part of your Umwelt along with the other important people in your life. It is what you know about her. It is the way you understand her. It is the model that allows you to deal with her.

Multi-modal representations are a key to understanding how the brain works. We cannot measure them. They are another mystery.

Popper's World and Dennett's Hierarchy

We now have internal models of the world and mechanisms for manipulating those models. What does this allow homo-sapiens to do? The answer is a great deal.

The philosopher, Daniel Dennett, proposed a hierarchy of abilities to respond to the outside world which he called generate-and-test levels:

The first level contains Darwinian creatures: These creatures are endowed by Darwinian evolution with a set of responses or tools. In response to signals from their environment, they generate responses and test their tools. Earlier, we called them replicating autonomous toolkits, and those with the best toolkits survive and reproduce.

The second level has Skinnerian creatures, named after the behavioral psychologist. (We named them after Hebb earlier.) They can improve their responses by learning but can only learn by conditioning. They blindly try out different tools on their environment; some cause bad results and are dropped; some are rewarded or positively reinforced, and the next time around are chosen first. Positive reinforcement is useful, but mostly if a useful response is picked, and only if the animal is not killed by one of its errors.

Craik's internal model of the world allows more. His model was more than a set of representations. It was a working model whose representations had a "relationship structure" with the object represented. Such a model could be manipulated to test possible actions and outcomes. It could predict.

Dennett's third level has such internal manipulation models. The creatures of this world try out model tools in a model environment map and pre-select the best one before they do anything. Dennett calls them Popperian creatures, after the philosopher, Karl Popper, who said that pre-selection "permits our hypotheses to die in our stead".

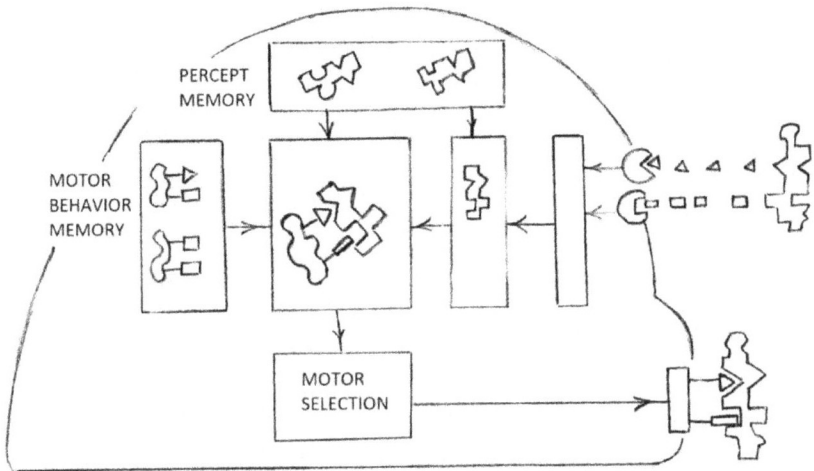

Figure 125 Internal Environment Manipulator Model: This is the same diagram as in the last chapter, but the motor behavior is now being manipulated to see how well it will perform with the percept.

Where is the map? Where is Craik's internal model? Almost certainly in the frontal lobes where all sensory areas are represented, where attention is directed, and where motor activity is planned and executed.

CHAPTER 12

HOMO CONCEPT USER AND CONCEPTUAL SENSE ORGANS AND DEMOCRITUS' WORLD

At neurology grand rounds, an attending physician would sometimes stop a junior resident in the middle of the diagnosis of a patient problem and say, "It's the swan."

There were no swans at grand rounds. The reference was to Zeus disguised as a swan lying on top of Leda. The uninformed resident might think that Leda was sexually perverse. The informed resident would know that the problem was with the swan. This was grand rounds code for a category mistake. The junior resident had diagnosed a problem with the brain when the problem was with the mind.

The swan has not been around forever.

About 80,000 years ago, death entered the world. We introduced it. We had changed. We had started to bury our dead. In so doing, we ritualized the loss of an individual. If we could ritualize that loss, then we could presumably conceive of an individual as a self and death as the end of a self.

The brain had not changed in size for some time. Some other change had taken place, and it seems likely that the ability to use the brain had improved to the point where it could contain concepts including one of self, and have a mind, and be able to think.

What is a mind and how do you think with it? At this point, I must admit to feeling out of my depth. I am not aware of an accepted working model of thought.

This is not completely foreign territory to neurology, but here we overlap with psychiatrists, psychologists, and philosophers. The philosophers have done much of the deep thinking, and with their help we are going to tackle the swan. We are going to see how well mental models explain the mind.

How do we model a thought?

This is not wholly foreign territory. A thought is a representation in an area of the brain. It is a mental model.

It is not a new thing to us. The setting of the olfactory bulb to an expectancy state can be thought of as a low-level thought about what is likely to be swimming around. It is not, however, an explicit thought. It has no symbolic representation. It is only implicit in the perceptual machinery.

The cortical area that sets the olfactory bulb expectancy state could have an explicit representation. It could have code for: reef-detected-so-set-receptors-for-reef-fish.

Thought representations must be, like perceptual representations, memory structures linked to sources of sensation in the outside world and informed by those sensations, but at a higher level of abstraction. This is a coding change, a new brain language, and an emergence.

What is the difference between a thought and a percept?

A major difference is that a thought is divorced from the machinery of perception and so a manufactured construction of the brain. It is an abstract representation. It can be manipulated and changed. It is so far abstracted from the mechanisms of any one particular brain that it can be shared with the brains of others.

Imagine you are heading up a shipping channel and hear a buoy sounding, beep...pause...beep...pause. You come into view of the buoy and see it flashing, red...pause...red...pause. You are not a multi-modal perceiver, so you cannot perceive the two signals as the same, but you can conceptually equate the two. You can manufacture the abstract concept of a signal followed by a pause. You can manipulate that thought and construct the idea of an on-off signal. You can tell others about both concepts. (After you have developed speech in the next chapter.)

How did thoughts come to be?

The most abstract manufactured constructions of the brain we have encountered so far are the multi-modal representations. A multi-modal sound-sight-touch is not a sensory but a symbolic representation of different stimuli, and an explicit symbol. These are likely candidates for thoughts, or for mental models that could evolve into thoughts.

What good are thoughts?

Thoughts allow us to further interpret the sensory input from the world outside. As a perceptual processor can recognize faces or fail to and register all faces as indistinguishable, a conceptual processor can recognize or fail to recognize a concept. Until you have read A. A. Milne's, The House at Pooh Corner, the word, "pooh", is an expostulation; after, Pooh is a bear of very little brain, and the phrase, "Pooh would say," means something.

The pre-Socratic, pre-Platonic philosopher, Democritus, said that a concept is like a sensory organ, a tool that allows interpretation of a signal from the outside world. A missing concept causes what could be called conceptual agnosia, and leaves the mind subject to, not misinterpretation, but missed interpretation of a sensory signal. Pooh is something other than a small bear.

Thoughts allow greater levels of abstraction like on-off signals. A perceptual system can generalize from presentations of various shades of blue to a general blue. A thought system can generalize further to the abstract concept of color. The thought system of Plato can move the abstraction up to the more general notion of quality. Plato was the first to use the concept and invented the word.

Thoughts allow new action in the world. Concepts are abstract enough to operate in both sensory and motor spheres. A tree branch pushing on another can be seen to be acting as a lever, which can open a new conceptual use for a stick. A concept can act as a motor organ. Richard Gregory first pointed out that concepts could be used as what he called mind tools.

Now we have three neurological metaphors: A concept is a sensory organ. A concept is a motor organ. And, from the last chapter, a behavior is a learned conceptual motor organ.

How do we deal with thoughts?

We will assume that we handle them, to a first approximation, like percepts:

We code them as mental models, but in a more abstract coding system.

We interpret sensory inputs as coming from objects in the world represented by mental models. We do conceptual pattern recognition.

We conceive self-generated mental model concepts including motor concepts. We manufacture conceptual patterns.

We consider or internally manipulate them.

We sometimes translate them into action.

A simple thought-handling model has sensory input (s) translated into a sensory concept in sensory concept space that can go to a concept manipulation space where it can interact with other concepts or can go directly to a motor concept space and be processed by an output stage into a brain output (x).

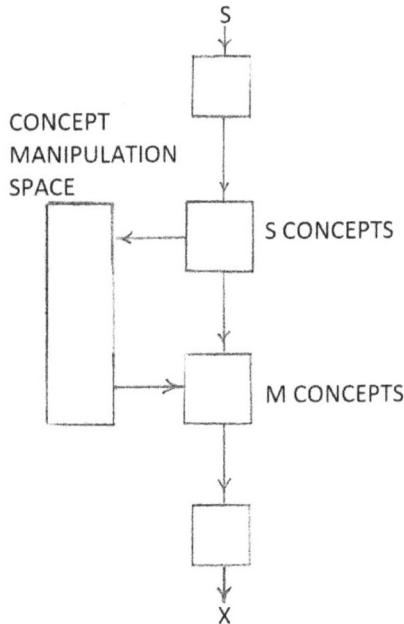

Figure 126 Concept Handling Model: This is like the percept-motor manipulation model of the last chapter. Input sensation is transformed not into percepts but sensory concepts in sensory concept space and then motor concept space or concept manipulation space.

How do we translate them into action?

In the last chapter we discussed Haugland's model brain run by committees of functional homunculi. The homunculi were thoughtless. Now we can entertain the idea of thoughtful homunculi generating thought sequences in concept space: If my girlfriend is home, then I will ask her out to a movie. If she is not, then I will go to the gym instead and do sit-ups until I have counted to one hundred. These statements are conditional switching statements and two of the basic computer programming instructions, the IF/THEN statement and the DO loop.

Young's idea of brain function as brain language handling seems to be appropriate for a computer brain. Is it for us? How much of human brain function could be explained by programable languages like this?

We will consider this later. First, simpler conceptual issues.

Pre-verbal Concepts

Most concepts and certainly complex concepts are language-based, but there are pre-language ones as well, like looking at the sky and thinking it looks like rain. We are going to deal with pre-verbal concepts for the rest of this chapter.

Where do such concepts come from?

David Hume thought that all concepts about the outer world came from sensory experience. If such concepts (in a book) could not be shown to come from sensory experience "...commit it to the flames: for it can contain nothing but sophistry and illusion".

He did allow another kind of concept derived from the operations of reason: A bachelor is unmarried. One bachelor plus one bachelor equals two bachelors. But there is no way to go from this kind of definitional concept to "this bachelor is tall" or "I see two bachelors". We must sense these things, so knowledge of the world can only come from sensory experience.

Emmanuel Kant disagreed. He thought there were three concepts about the world prior to experience: space, time, and causality. He thought these were built-in, conceptual spectacles through which we interpret the world.

We said earlier that registering space and time was part of having a hippocampus, so we seem to agree with Kant. But are they part of the conceptual apparatus or implicit in the perceptual apparatus? They are certainly concepts but seem to operate at the other level as well. This is not unique: The color, red, is a receptor response, a perceptual "decision" in the absence of green, and a verbal concept.

Karl Popper disagreed also. He thought that a theory of what was available to sense had to come before sensing, and that sensation was then filtered through these "theoretical propensities" for interpreting the outside world. Space and time were such propensities built into the nervous system. His concept of theoretical propensities included the sensory receptors which he considered "theories" about what stimuli were available. These "theories" were constructed by evolutionary experimentation.

Neural networks have propensities. Our examples had larger responses to certain inputs. Every matrix and so every network has what are called characteristic functions or inputs that it handles in special way, passing them through unchanged. These special functions are determined by the network numerical structure. They are properties of the matrix. Since the matrix can learn and change, these functions and network propensities can learn and change as well.

Propensities and the Bayesian Brain Hypothesis

Space seems more perceptual than conceptual. Its boundaries are tangible. We live in it and move through it. This three-dimensional reference frame seems impossible to do without. Without it, the objects we sense would be located nowhere. Can we imagine nowhere? Can we have a mental image of an object nowhere and in nothing? It seems impossible or at least inconceivable. Space seems mostly perceptual and built into the wiring.

Time seems more conceptual than perceptual. It is intangible. We can conceive of its absence. We can imagine time stopping. Movies represent it as motionless, unchanging space. We can imagine animals or infants living in a continuous present without awareness of it. Time seems to be more conceptual and to require experience and memory to appreciate, but it is also part of the perceptual apparatus and memory lets us perceive its passage.

Causality seems even more conceptual. It is intangible. We do not sense it. Hume denied its possibility. The generative mental model, however, suggests that it could be learned from experience.

The Bayesian brain hypothesis is a probabilistic version of the predictive coding version of the Craik hypothesis. Bayes Rule is a way of dealing formally with the fairly obvious fact that, if two things often happen together, one can predict the other: a set of hoofprints, suggests that a horse has been around.

It can also deal with changed probabilities due to contexts such as environment: A horse is likely in Texas; a zebra in Africa. It can update the probabilities after new evidence: Horses and hoofprints earlier today increases the probability of one predicting the other now.

Such a Bayesian brain could conclude that hoofprints predicting horses means that horses cause hoofprints even if it did not see this happen. It could equate correlation with causation.

It could confuse the two as we all still do. After ten sunrises, a Bayesian brain might predict a near-certain probability of sunrise given a pink sky, but it might also calculate that the earlier pink sky caused the sunrise. Rosy fingered dawn might not be just a Homeric cliché but the goddess of the dawn poking Apollo to get up and do his job of rolling the sun across the sky. The Bayesian prediction could be right even though the mental model of causation and the physics were all wrong.

A Bayesian brain could create a Popperian propensity to manufacture a Craikian mental model of causality.

Maybe. Regardless of how they work and what they are exactly, Kant's spectacles and Popper's propensities seem to exist and to help us interpret the world. They allow us to build our mental models of the world.

Speculations on Mental Models of World and Self

What are the earliest conceptual mental models like?

Birth requires an infant to become aware of outside space and make a first distinction between itself and the outside world: Self and Other. Self and Universe. Me and It.

Figure 127 ME and IT

It probably does not do this well. The infantile nervous system is only partially myelinated. Its sensory systems do not work well. The sensed outer world must be chaotic—"a blooming, buzzing confusion", as William James said. There are the pre-wired propensities, including a preference for looking at human faces, but perception must be minimal, and comprehension impossible. The sensed body, similarly, can be only a confusing notion. Its mouth and feeding sensations are probably the most clear—like the tunicate. These initial representations of outer world and self are bad models, but they get better.

Jean Piaget thought that the first two years of life were devoted to learning sensation and action: learning to perceive the outside world and building mental models of things in it; learning to perceive the body and building a mental model of it; learning how to do things to the world with the body and building motor mental models. (Piaget did not use the mental model theory, but his concepts can be fitted into this framework.)

He thought the major achievement of this stage was object permanence or learning that objects still existed when they could not be sensed. This would allow continuity of perception of the outside world. The mechanism in our brain model would be maintaining memory representations in the internal world model. This would create some sense of time at the perceptual level.

Knowledge of the world and body develop together. Perception and action develop together. As limb control develops, the touch map incorporates the arms and legs into the body model. Like the man in the box, the limb sensors can then be used to build a map of the space around, usually a crib. To explore the crib further, the infant can grasp the bars and stand, which changes the perception and the map. This map is close in and egocentric. Later the outside world moves to the limits of the distance sensors and becomes less so.

When vision starts to work, the world really opens up, but perception and action still develop together. We discussed investigators who sewed eyes shut. In a more-subtle experiment in 1963, Richard Held and Alan Hein yoked two kittens together so that one could move freely while one sat in a basket and was moved by the other. Although they had the same visual experience, the basket kitten did not develop normal visual processing areas. The only difference was passivity. We seem to have to act in and on the world to learn to properly sense it.

William Ashby, in his book, Design for a Brain, suggested that moving through the world is like changing the world by your actions: you pull the world ahead toward you and push the world around toward the rear. This rather odd conceptualization may explain how the brain sensory systems "think" about movement and why the kitten could not see: If sensation changes, it can be because of change in the outer world or change caused by your movement. To separate the two, movement-induced sensory change must be learned as consistent with body activity: No activity, no learning. The hippocampal map, where body movement is represented, is likely to be important for this sort of thing.

When the infant escapes from the crib or the basket and starts crawling around the house, it must use the hippocampal space map, but it does not use it well. Piaget observed that it takes two years to develop a map that incorporates the ability to go out of the kitchen into the living room and around to the other door and back into the kitchen. We may have spectacles for space, but it takes us a while to learn to use them.

We sense and move and calibrate the models. We continue to calibrate and refine them for the rest of our lives. The world and the self are always a work in progress.

Where is the Self?

Where is this self? Where are "we"? We have areas that do this and areas that do that, but where are "we", where are our souls?

This is not a respectable question in neurology. It is respectable, however, to ask where is the self-representation?

It is clearly in the brain or at least in the nervous system. Francis Crick thought Alice might have said that she was "just a bag of neurons". In the bag there are multiple representations of the self: in the spinal cord, in the brainstem tectum, in the thalamus, in the hippocampus, in the cortex touch maps, and in the frontal lobes.

There are pathologies of self-representation:

The self can lose bits of itself: Patients with tactile neglect can become unaware of parts of their bodies.

The self can register false bits of itself: Patients with amputated limbs can sense phantom limbs, parts of themselves that are no longer there.

The self can be persuaded that a foreign object is part of itself. In the rubber hand illusion, the self's hand is hidden by a screen but beside it is a visible rubber hand. The experimenter strokes both hands at the same time, and the self seems to feel the touch on the visible rubber hand, not its own.

These are pathologies of the body-self. The body-self seems to be a mental construct that breaks down if the nervous system is damaged or the experimental situation distorts reality too much.

This is not the interesting question though. Where is the real self? Where are "we" really? Where is the essential mental being, the me-self, as William James called it?

Descartes was the first to have a go. He was interested in the soul rather than the me-self, and thought it was in the pineal gland. This seems odd but there was a rationale for it: The theological soul was a unitary entity. Aristotle's common sense was unitary. Since the pineal is the only unpaired structure in the brain, it had to be in there.

The theological framework seems to have driven Descartes' thinking in an odd direction. There have been more plausible candidates since.

The hippocampal episodic memory map is one. You are part of the episodic memory of your going to the beach last summer. This is to be distinguished from semantic memory which contains what you have learned about beaches. Episodic memory contains a self. You are part of that memory and part of a self is its memory of its own continuity through time. The patient HM with the damaged hippocampus did not lose his sense of self; he lost the future of his self. The hippocampus is necessary for changes to the self but is not the thing itself.

The multi-modal maps are candidates. There are multi-modal possible selves in the tectum, the thalamus, the sensory cortex, the claustrum, as Crick and von Malsberg thought, or the frontal lobes where multi-modal sensation goes and where the thalamic sensory streams are selected.

The frontal lobe multi-modal areas are particularly attractive candidates. Strokes there can cause dramatic alterations in personality. The most closely held ideas of a personality are concepts of gods and sexual behavior and with frontal lobe disease they can change dramatically: The pious can become irreligious; the chaste, promiscuous. We noted earlier the devastating changes of the classic frontal lobe syndromes: total apathy on the one hand, and the disinhibited state that was "...not the same Gage..." on the other.

There are other pathologies of the me-self.

It can be split. Patients can have the two halves of their brains disconnected by severing the connection pathways. (I am just going to pass over why neurosurgeons do this.) Roger Sperry tested such patients in the 1960's and their selves seem unchanged. They noticed nothing odd about their disconnected states and seemed quite normal. Only if they were put in a test situation where the sensory inputs to the two halves of the brain were separated, did it become apparent that there was something wrong: Such a young woman blushed when a photo of a naked man was displayed in her left visual field and thus right hemisphere, but could not say why. Her me-self seemed to be in her left hemisphere.

These patients can develop a bizarre condition called the alien hand syndrome: The left hand is controlled by the disconnected right hemisphere. It can escape volitional control and reach out and grasp objects on its own. The self in the left hemisphere shouts at the hand to stop.

Is this how we are to think about split-brain patients? Is the me-self in the left hemisphere? Has the hand escaped the self? Or do these patients have two selves?

Can the self be obliterated? Can there be complete loss of the me-self? What about amnesia patients who forget who they are?

They do not exist, or, rather, they exist only in movies. Rarely patients do complain that they have lost their sense of identity, but this is a psychological condition, a reaction to overwhelming psychological stress called a dissociative fugue state. The sense of identity returns.

Patients like H.M. can lose large parts of their lives, but never their complete sense of self. On the other hand, partial loss of the me-self occurs every day. Where is the ten-year-old you?

If the me-self has a specific location, strokes should reveal it. They do not. Strokes in the frontal lobes do not obliterate the self; they diminish it, and strokes in other areas do as well. What I see, day after working day, as strokes destroy parts of the brain and as neurons wither away in patients with Alzheimer's dementia, is the fading away of the me-self. My patients become less able to experience the world and less able to act in it. They become less able to access their memories and less able to think about them. My clinical job suggests that the me-self is a distributed representation and can suffer graceful degradation.

Popper said that we learn to be selves. We accumulate memories of what we have experienced and learned from experience, of what we have thought and learned to think, of what we have done and learned to do, of our interactions with other selves and what we have learned about selves.

The me-self seems to be both distributed and learned, and more of a process than a thing.

Speculative Infant Psychopathology

We can speculate about infantile me-self psychopathology. Freud started this sort of thing. He was interested in the effects of infantile sexuality. We are more interested in the effects of the infantile mental models. Since these models are speculative, these are speculations built upon speculations. Not testable hypotheses to be relied upon, but fun to think about.

The infant can first be aware only of Me and Other, and Other is the "blooming, buzzing confusion" outside. That is all there is. Everything that happens, happens to the infant. It is all about him.

This infantile egocentricity may underlie our later tendency toward self-referential thinking: the teenager's certainty that everyone at the dance is staring at him; the adult's suspicion that the IRS is after him. These are delusions of self-reference. We all have them: Doesn't the honking horn in the car behind always seem to be directed at you?

Self-referential thinking is a simplistic thinking style in which everything is explained by the universe's direct interest in the thinker. Paranoid thinking makes the interest inimical. It is characteristic of psychotic states like schizophrenia, but a neurologist sees it most often in delirium and dementia. It explains everything—easily: They did it! They're after me!

The infant self, having made its first distinction between Self and Other, must then develop a concept of The Other.

Figure 128

The Other sings lullabies to us. She picks us up and feeds us. She acts. She is volitional. She is all that happens outside us. She is omnipresent. She is the universe.

We can talk to her. She comes when we cry—sometimes. We can control her. She does what we want—sometimes. Some of the things we do work, and others do not. She is omnipotent but capricious. She is scary.

Could anything be more egocentric than the notion that the universe is an entity responsive to our every thought and act? Does dealing with such a universe predispose us to religious thinking?

Were the early gods memories of the infantile Mother-Other? An early godless psychiatrist would diagnose the Olympian Gods as delusions and religion a shared delusional system.

Does this account for another simplistic thinking style, the tendency to project a volitional agent into the outer world as the cause of everything? Zeus, and Thor in the Norse pantheon, cause lightning and thunder. They-did-it conspiracy theories explain everything. Does Mother always do it?

Knowing what is going on in minds is not neurology but psychiatry, the study of the psyche. It's the swan. Psyche was a goddess (and you remember who the swan was), and the practice of psychiatry was thought by the philosopher, Paul Recoeur, to be like dealing with such entities. He called it a branch of hermeneutics, the study of oracles.

Hermes is the god of messages and oracles. His sibyl breathes the funny gases coming up through the temple floor until she is drunk enough for the god to enter her and speak. She inspires and is inspired by the god.

The priest tries to figure out what the god is saying. The modern priest consults his divinely-inspired book and tries to figure out what the god is saying. The psychiatrist listens to the patient talk and tries to figure out what the hidden and subconscious areas of the mind are saying.

The psychiatrist has an easier job than the priest. The priest is dealing with an alien mind. The psychiatrist has a model of a mind, his own, but it is still difficult to figure out what is going on in the swan.

Other Selves and Other Minds

To go on to a more concrete topic, the misty All-Mother eventually resolves into an outer world with a mother-self in it. To deal with another self, the infant needs to develop a theory of how other selves operate.

Figure 129

Other selves have minds that direct them. To deal with other selves with minds, the infant needs a model of how they work, a way to think about what others are thinking, a theory of other minds.

Figure 130

This is a new kind of entity in the internal world model, another mind. The infant can preview and pre-select its ways of dealing with it, but it takes the infant, if not the adult, quite a while to pre-select well.

Not all infants can do this. Autistic infants are not cuddly. They do not prefer to look at faces. They do not smile or respond to smiles. When grown up, they are odd. They have compulsive behaviors. They find much human behavior strange. They do not get it. They cannot tell if people are lying to them. One gifted autistic described her situation as like being an anthropologist on Mars, in Oliver Sack's book of that title. She lacked a theory of other minds.

The rest of us do and use it all the time. We look at faces and guess about feelings; we look at actions and guess about thoughts; we listen to words and wonder what was really meant by them. Human communication is dependent on an internal model of the insides of other peoples' heads, a concept of mind.

Self-Consciousness and the Me-Self

The self can make a mental model of itself. The self can regard the self.

But not at first. Although we have said that infants develop an early concept of self, this does not mean that they are self-aware. Their first self-concept would be more like a nebulous feeling of distinctness than a true concept. As the mental structures become more sophisticated, so does the self-concept. Once again, Karl Popper said that we learn to be selves.

If you paint red dots on the noses of infants and put them in front of a mirror, they ignore the dot. Only at the age of two, do they connect the dot on the face in the mirror with themselves and wipe it off. The interpretation of this experiment is that they have become aware of themselves as selves; they have become self-conscious.

Figure 131

Whatever else self-consciousness entails at that age, the clearest thing about this test is that children can see themselves as objects in the mirror—and in the world. They have moved from a subjective stance to an objective one.

In our model of concept structures, self-consciousness would seem to require a self-symbol inserted into the model of the outer world, an objective model, a me-self. Now it can think about itself as well as you.

The me-self can go on to elaborate a mental model of its own mind. It can regard its own internal workings. It can think about what it is thinking—and what you are.

It can move on to more dizzying levels of mental simulation: It can think about, not only about what it thinks, but what it thinks you think, or even what it thinks you think it thinks—and plan for it.

Self and World Models and Understanding

The outer world takes shape as containing other selves with other minds.

Figure 132

We are good at dealing with other selves and minds. We effortlessly carry out theory-of-mind operations. Those operations are performed using mental models of what is going on in other peoples' heads. The models show us how things work. We project those operations out into the world of other selves and are usually right.

The outer world takes shape as containing entities other than selves. Some are volitional entities like simple selves, and some are purely physical entities that act only according to the laws of physics.

Figure 133

We are good at dealing with the world. We effortlessly carry out theory-of-world operations. We may start with misconceptions of volitional agency (That branch tripped me!), but we correct them, and our models show us how things work. We project these operations out into the world and are usually right.

We understand how a self or a thing in the world works: We conceive a model that bears a "relationship-structure" to the corresponding object in the world and develop it until it "...works in the same way as the process it parallels". We then understand that object, which is to say we understand how it works, and can predict what it will do.

First using the non-verbal concept space.

Figure 134 Mental model of the contents of mother's mind.

Later the verbal and even scientific concept space.

Figure 135 Mental model of the forces acting on a bridge.

This is my understanding of the theory of understanding of the psychologist, Phillip Johnson-Laird, translated into Kenneth Craik's terminology. Such understanding would seem to depend on generative mental models like those of complex perception.

Craik pointed out that explanation of any sort allows prediction. This model of understanding allows prediction. It could be useful and selected for by evolution.

The task of the developing mind is to learn to master this mental model world in its multi-level complexity: To leave the egocentric world of infancy and move up into the more abstract reaches of the cortical map. To understand what is happening in the world. To consider action before acting. To predict outcomes. To learn to use the instrument in the head.

To do this it must suppress its older, emotion-driven responses. It must use its frontal lobes to master the older responses to allow the instrument in the head time to work. It takes a long time to master the instrument, and a long time to grow up.

The Processing Perspective on Self

Can we make a thinking machine?

We could now take the model of the last chapter and try to convert it to a thinking system. We could add an abstract representation level. We could design symbolic representations and a mechanism for manipulating those symbols.

This would be difficult to do. How would we design a thought symbol? What would motivate the manipulation of those symbols? Would there have to be a level above that to tell it what to do—and then one above that?

The connectionist neural network model makes one approach easy: Run an experiment. Add another level, turn it on and see what happens. The model would learn, perhaps it would learn to think.

It seems unlikely that this would work, but it might. How would we know if it did? How could we tell if our model was thinking? How could we test for intelligence? How could we know if it had a Self?

Alan Turning addressed this question for computing with what is now called the Turing test: Put a computer in one room and Turing in another and pass messages from room to room. If he cannot tell that the computer is not intelligent, the computer is intelligent.

We could do another connectionist experiment: We could feed self-related signals to a multi-modal map to see if it developed a self-symbol. Again, how would we know? What would one of those look like? Would Turing be able to solve the problem? If he cannot tell it is not a person, is it a person?

This may be beginning to seem like mad scientist territory, but Turing's test is a reasonable approach. There is a problem. This is not testable science. Turing is doing hermeneutics: He is looking at outputs and try to guess what is going on inside.

Does he have any other choice? Could he devise an objective test for conscious intelligence? Could he formally measure a Self? Is he not forced to look at speech and behavior outputs and guess, whether he is assessing a computer or meeting you?

The Self and Self-Programming

Beyond the primitive drives of the amygdala, where are the motivations that drive the Self?

The answer is: Within the Self.
What manufactures those motivations?
The answer is: The Self does.

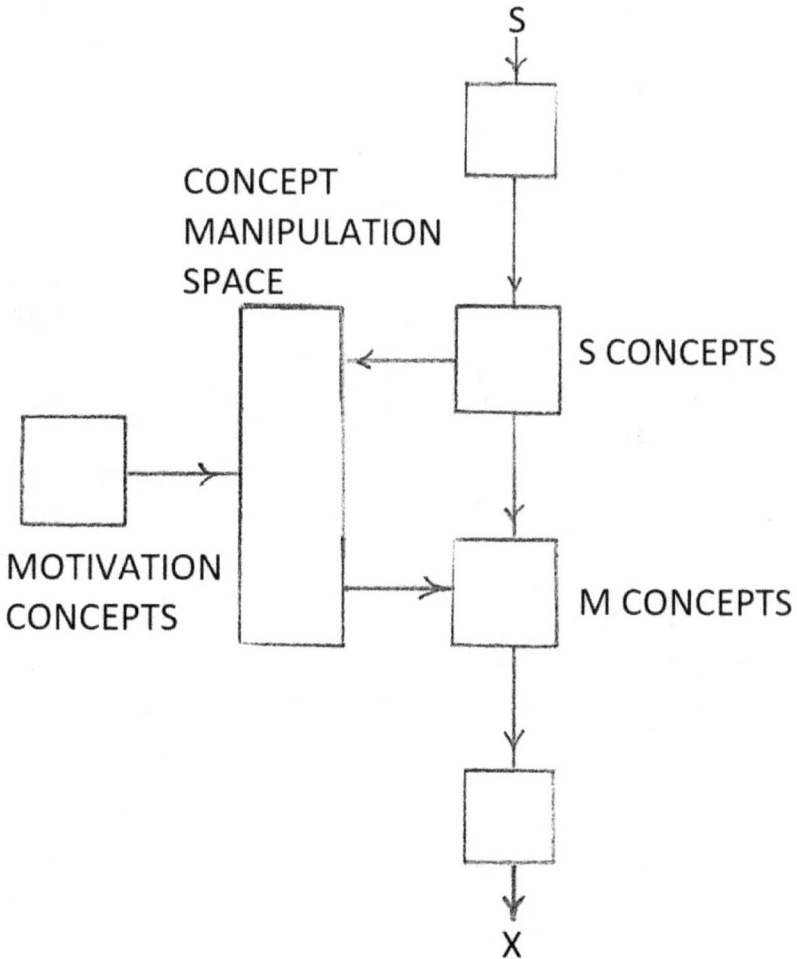

Figure 136 The conceptual motivation space.

Can we make a model that creates its own motivations? How do we make a model that makes itself want to do anything? The answer is: We make a self-programming Self-computer.

The self-programming computer is not a new idea. The 1936 Turing machine was a theoretical universal computer with an infinite memory that could manipulate its own memory of instructions. The

1945 von Neuman machine was a real computer with a program of instructions stored in a finite memory that could manipulate data but also treat its own program as data and manipulate its own instruction set. It could change its own goals.

It has been suggested that we are the same. If so, the Self becomes a programmable process or function of the brain: Modifiable software, not hardware. A disturbance, not a fixed part. Not only distributed and learned, but somehow immaterial.

One or two finite state automatons did not seem adequate to explain the mind, but Haugland's hierarchy of many automatons was plausible.

Finite state automatons are not only machines but language processors. They figure in the hierarchy of computer-language processors. They are the limited processors at the bottom level. At the top level is the Turing machine.

If we make the high-level homunculi programable, language-processing computers that can modify their programs, will that do to explain the mind? Some say, yes, that is all we need. The brain is a computer. The mind is a Turing machine. The Self is a subroutine. This is an ongoing debate.

The Self and Free Will

The Self seems to be a mechanism for the brain to exercise free-will, but now we run into a philosophical problem: Is volition completely determined by inputs and the prior state of the Self? If so, it is pre-determined and there is no free-will. If not, the Self must inject novelty. If truly novel and not determined by prior internal states, it must be imposed or random and so not in the control of the Self.

Our model has moved the freewill problem into a perception-action-representation-matching theatre where putative volitions can be previewed by a self-computer which then selects an action. Is this free-will?

The self-programming Self computer could change the action programs. If so, would it do so on the basis of its prior states, or would it inject novelty? If the latter, where would novelty come from?

Libet's experiment suggests that our apparent volitions are the result of something like network learning, of long chains of ΔB-modifications of the brain matrix. That they are predictable and only falsely experienced as self-determined. That Libet could write the equations for them if he had all the data.

Could the neural network model help with the free will problem? Free-will seems to require novelty. Could a neural network create novelty?

Leon Cooper wondered if a network mistake could. Noise in the system could let a stimulus evoke the wrong output and process an r'-input incorrectly as,

$$fm = M. \, r'$$

$$= (\Delta M' + \Delta M'' + ...). \, r'$$

$$= f''$$

which could be thought of not only as a novel response but a network metaphor: This is that.

Could such errors move thought into random and novel grooves, as gene copying mistakes do? Or are such mistakes predictable? Will some future mathematician be able to write the equations for them?

Have we solved the problem or made it worse? In a discussion of this problem, Marvin Gardner quoted Piet Hein to suggest that the problem is like

"...two locked boxes. Each
Contains the other's key."

This may be the final word on the subject, but I have an opinion: I think that our brains can make new associations such as metaphors and analogies, and so generate novel concepts: I think that we can rub two strands of the material neural world together and make conceptual fire. If so, the development of human thought is not predictable but a chance-driven, groove-shifting, random process. If so, this is grand but there might be costs.

Every step forward in representational complexity seems to generate new pathology. What about the Self-representation or Self-function? How can a self-modifying mental representation go wrong? What sort of trouble can a self-programming Self-computer get into? I cannot even begin to imagine—but I do not have to—I can just look in the mirror.

Democritus's World: The Human Mind and Its Conceptual Organs

The human mind contains an internal world populated by abstract representations of the outer world called concepts. Concepts are mental models so abstract that they can act as sense organs or motor organs. They are symbols and can be freely manipulated. They can be prediction organs. They can be motivations and drive behavioral motor organs.

This world is Craikian, Popperian, and Democritian. In it, concepts and conceptual actions can be manipulated and tested. It could be considered a seventh functional part of the nervous system.

The concept world has conceptual sense organs that allow the interpretation of signals from the outside world that would otherwise be meaningless. It has similar conceptual sense organs for what is going on in other peoples' heads. It has a conceptual model of itself which also functions as a conceptual sense organ for what is going on in its own head. That model can be changed. It can be changed by itself, by what is going on in its own head.

Conceptual models and such sense organs are tricky. They require careful handling. They allow new kinds of mistakes.

Figure 137 The phantom triangle. (After Kanizsa, 1955.)

Is the phantom triangle a mistake? It does not exist on the paper. Gertrude Stein might say it is not really there there. She might say it is a metaphysical, not a physical, entity.

It does exist in the nervous system as a set of neuronal impulse trains. It is a physical entity there.

It is a percept, a shape, a neuronal construction of a perceptual sense organ. Much of what we perceive as real is like that.

It is a concept, a triangle, a neuronal construction of a conceptual sense organ. Much of what we conceive of as real is like that.

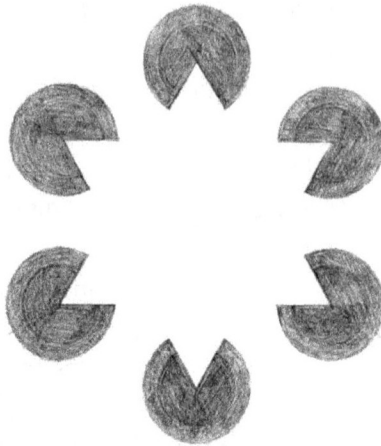

Figure 138 Another phantom.

What about this drawing? Is it two intersecting triangles or a single polygonal star? Is it one shape or two or six?

Does it take us further up the conceptual ladder? If the many shapes evoke one, does it evoke the phrase and concept, "e pluribus unum"?

We look at faces and conceive of feelings. We look at daring behaviors and conceive of bravery. We look down at boards and conceive of bridges that bear our weight.

We look up and can conceive of Sun Gods or conceive of thermonuclear reactions. How do we decide what is true? How do we test our conceptual hypotheses for truth?

Carefully! Popper's criterion was not truth but falsifiability. If you deny that you stole my money, I can check by looking in your pockets.

If I find nothing, I have not proved a truth, since the money could be hidden elsewhere; but if I find it, I have conclusively proved a falsehood.

Popper thought truth was unachievable since any new fact could undermine apparent truth. He thought falsification was solid and could not be undermined.

If you say the devil made you do it or you did it because your mother did not love you, there is no way that can be proved false or true. Popper thought such concepts could not be falsified. He did not say they were not the truth, but only that they were not testable scientific truth. Psychoanalysis is not testable science. Neither is Turing's test.

Truth is more elusive than falsifiability. Concepts require careful handling. They are tricky.

Concepts differ from percepts in one important way: They can, with care, be made perfect, or at least close to perfect. We cannot see past our receptor sets and perceptual modules, but our conceptual structures can. They can overcome the limitations of the sensory input systems. They can be refined until they describe the world—indeed, the universe—as it really is. The Umwelt can be transcended. We can escape Plato's cave.

Only if we do it well. We can handle concepts badly and become deluded. We can handle them well and become wise. The philosopher, Georg Hegel, is said to have said that we, both as individuals and as a species, refine our internal concepts to strive for exact correspondence with and perfect apprehension of the outer world. He also thought that, as concepts became more exact, poor responses would be eliminated until, with perfect apprehension, only perfect responses would remain.

This does not seem to have happened yet, and Popper would say it never will, although we can get close. It is certainly something to aspire to.

CHAPTER 13

HOMO LANGUAGE USER AND SEMANTIC MIND TOOLS AND GREGORY'S WORLD

About 40,000 years ago, man began to draw animals on the walls of caves and chiseled images of the phases of the moon on bones. These representations are visible thoughts. They are also broadcasted thoughts or communications.

Primates communicate with noises. As they move through the trees, they chatter to maintain contact. They make noises that convey emotion like "Ouch", and warnings like "Watch out". The vervet monkeys have the most warning cries: one for danger on the ground, one for danger in the trees, and one for snakes. These are stereotyped noises with fixed meanings and no way to say anything new.

We use descriptive speech which is unique to us and quite different. It has a theoretical mechanism called semantics that gives meaning to and extracts meaning from noises, and a theoretical combination mechanism called syntax that codes and decodes lawful word combinations and relations. This combination system can produce an infinite number of novel sentences and these sentences can be disassembled and understood by any language-competent listener. It is a coding change and an emergent property. It is so far beyond the capabilities of animal communication that it is another leap of the and-now-a-miracle-occurs sort.

It is not innate, although there are propensities. It takes two years before our brains can begin to use even simple concrete speech, and this marks Piaget's second stage of cognitive development.

Piaget thought that a major feature of this stage is the ability to pretend or construct hypothetical narratives: If I were grown up, I could drive a car. What would happen if I took the keys and did that? I probably shouldn't tell! This is concept manipulation in the verbal sphere. It would seem to underlie predicting and planning—and lying.

Speech is a complicated subject with a mind-numbing literature. I am going to keep it simple and confine myself to a few clear points.

The Neurologist's Simple Speech Circuit and Anomic Aphasia

Neurologists can usually get by with a simplified speech model with a speech recognition area in the left temporal lobe first described by Karl Wernicke and speech production area in the left frontal lobe first described by Paul Broca, both in the nineteenth century. They are connected by a fiber bundle. Damage to the output area leaves the patient unable to speak; damage to the recognition area leaves a patient unable to recognize words but able to produce them in a scrambled, nonsensical way—a word salad. Damage to the connection leaves the patient able to understand but unable to repeat what is said. The blood supply to these areas is discrete enough that the neurologist's probe, the stroke, often damages just one of these areas and demonstrates its syndrome.

Aphasia (no-speech) is the neurological term: Wernicke's receptive aphasia for bad reception, and Broca's expressive aphasia for bad expression. Speech is more complex than this, but the neurologist can usually get by with the two-stage model.

The more interesting aspects of speech are the deep connections to memory and meaning. Ludwig Lichtheim in 1885 proposed connections to an area where concepts were located, and meanings determined. Wernicke's area would do sound recognition and this the deeper aspects. He later decided that it was more complicated and there were several centers. The later finding that man-made object representations are in frontal rather than posterior sensory areas fits with this idea.

A rare kind of stroke confirms Lichtheim's hypothesis by damaging the areas around and isolating the speech circuit. It leaves the patient able to hear and repeat, but with no comprehension of what has been said or repeated. The patient can handle the sounds but has no access to the meanings.

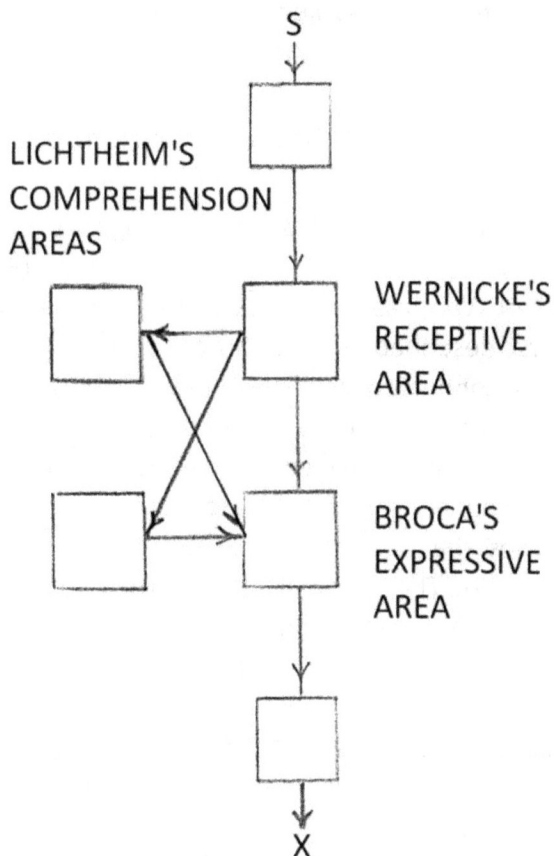

S

LICHTHEIM'S
COMPREHENSION
AREAS

WERNICKE'S
RECEPTIVE
AREA

BROCA'S
EXPRESSIVE
AREA

X

Figure 139 The speech circuit with Wernicke's receptive area in the temporal lobe and Broca's expressive area in the frontal lobe and two Lichtheim comprehension areas to the left.

The deep connections result in a number of deep aphasias. One such is the word finding difficulty known as anomic (no name) aphasia. Patients can receive and express, and know what they want to say, but cannot find the right word. They can remember everything but the name.

Word finding is a fundamental speech function. Every aphasic has some trouble with it, and it is often the last stage in recovery from any clinical aphasia. It is also a problem for normal people. Who has not suffered from the tip-of-the-tongue problem? Who has not forgotten the name of a person, or the name of a subject like anomic aphasia?

Names are more abstract than their underlying memories. They act as a filing system. They can be disconnected from the content they designate.

The Syntax Mechanism

Grammar book syntax shows us how to analyze speech: subject before predicate, and so on. It allows us to untangle sentences like: "The father of the boy who owned the baseball bat threw the ball."

It takes a moment to do that. The meaning is not immediately clear. Once done, however, two relationships and an action and a story are understood. There is more there than the words themselves.

Deep syntax was first proposed by Noam Chomsky in the 1950's and was thought to be an innate human brain mechanism. The theory is complicated and has changed over the years. A simple version translated into our mental model language is that deep syntax is an internal model, and comprehension is like generative perception: a spoken sentence maps onto deep representations in the model.

Syntactic speech must be determined by the deep syntax propensity and the intrinsic brain representation languages, but there must be more—some aspect that has developed over time. Derek Bickerton pointed out that there is a puzzle about human development: We went through three million years of brain enlargement without much change in our abilities. Our brains reached the lower limit of their present size one and a quarter million years ago, but we continued to "...squat in smoky caves and make the same stone tools." Then, forty thousand years ago, we suddenly changed and rapidly became the dominant species. What happened?

Bickerton thinks that syntactic language happened. He suggests that the earlier brain enlargement was driven by the development of a verbal representation system requiring a larger and larger brain. However, during this long period, speech never advanced beyond proto-speech like, "See banana. Me eat", which he calls the verbal level of "two-year-olds and signing chimpanzees". Then, forty thousand years ago, syntax emerged, and we had an advanced speech system. This allowed more advanced thinking and opened a new chapter in evolution. As he put it, " We blundered into language, and then became smart." Syntax was the driver.

How did syntax come to be? Did the brain change or did the mental model improve through interactions and learning?

Learning obviously plays a part in human speech development. Neural networks can learn to make speech maps. Helga Ritter used a self-organizing network to make a simple one. It was fed short sentences like, "Jim speaks well", and it dissected the sentences into nouns, verbs and adverbs and mapped them into separate areas in memory space. Speech was transformed into a topographic map. It was a see-banana-me-eat map. The experiment of childhood we all go through trains us with more complex sentences and produces a more complex map.

Did Chomsky's innate propensity need Bickerton's one-and-a-quarter-million years to learn syntax? This is a long time for a learning effect. Did the brain change in some way we cannot measure?

Syntax and Meaning

There is a curious thing about the syntactic machinery: There can be lawful sentences that make no sense. Chomsky's example was: "Colorless green ideas sleep furiously." The sentence is meaningless and some of its word relationships are impossible and yet the sentence still can be understood in a sense: Something impossible does something impossibly.

The sentence relationships are clear even if the word relationships are wrong. There is meaning inherent in syntax that is separate from the meaning inherent in words.

Chomsky's sentence is an example of syntactically right but semantically wrong speech. It could be produced by a formal, probabilistic, speech production system like a computer. Such a system processes its inputs and produces its outputs in a lawful, syntactical fashion without regard for meaning. Any meaning is in the mind of the human user.

No human being would ever produce Chomsky's sentence. Even a Wernicke receptive aphasic would not say it. There would be more meaning and less-than-perfect syntax. Human syntax works along with semantics. Lila Gleitman found that verbs with similar meanings use similar syntactic sentence structures or, as she put it, "...verbs of a feather flock together". Bickerton's blunder may have been mostly syntax but involved semantics as well.

Semantic Memory and Meaning

What about semantics? How do the implicit signals in arbitrary sounds become realized as explicit meanings in another mind? How do we code and decode?

It does not seem that difficult. If words are arbitrary, then they must be learned. If they are learned, we know how that is done: I point at a stick and grunt; you point and repeat. We train one another. The vervet monkeys are on the way: They have separate cries for danger on the ground and in the air and from snakes. Are these not primitive nouns? How long will it take them to get to, "See snake, me run"?

Once these sounds are learned, then word comprehension becomes assigning auditory inputs to the correct cortical representations in the speech symbol maps, the Lichtheim comprehension centers. This is a procedure no different from percept assignment, but at a later stage of the sound interpretation system.

We can use our content addressing equation for word meaning retrieval:

$$fm = M \cdot r'$$

$$= (\Delta M' + \Delta M'' + \ldots \text{etc.}) \cdot r'$$

$$= \Delta M' \cdot r'$$

$$= f'$$

This makes nouns seem simple, but they are not. If we agree that the word stick means a branch of a certain size, then we can communicate about sticks, but much more has happened in our minds. We have linked the arbitrary sounds to other patterns. Some of them are simple and definitional, such as visual appearance. Others are conceptual: A stick can be a club, a hammer, a spear; it can be firewood, or the framework of a hut, or something to lift a rock.

The conceptual semantic ramifications are endless. Once we have mentally linked stick to lever, it is a concept with the power to build a cathedral. Once we have linked it to a model of the planetary system, we can speculate with Archimedes about moving the world. The word becomes part of a map that reflects the life experience recorded in our brains. It becomes part of our internal model of the world and our understanding of the world.

Nouns are abstract designators of conceptual depths. They are like the visible tips of icebergs; the real structure lies beneath, and this deep linkage is where meaning lies. It is a web of perceptual and conceptual maps. It grows with experience and becomes an increasingly complex structure of inter-digitated webs where the stick web meshes with those for firewood and levers and astrophysics.

Henry James said: "Experience is never limited, and it is never complete; it's an immense sensibility, a kind of huge spider's web of silken threads suspended in the chambers of consciousness and catching every airborne particle in its tissue. It is the very atmosphere of the mind."

The web is called semantic memory and is a system distinct from the episodic memory system. Episodic memory is me-memory: I came. I saw. I did. Semantic memory is everything else. It is where we store what we know. It is the coded representation map of all we have learned—about everything.

The two may be the same. It has been argued that semantic memory is abstracted from episodic memory, that episodic memories can lose specificity and become semantic.

Words are both a part of this memory and its indexing mechanism. There is a disease of words and semantic memory, a rare condition called semantic variant primary progressive aphasia. It is not just an aphasia. It does begin that way as inability to recall names, but then progresses to loss of memory for what things are or can do, and later to generalized brain dysfunction. It is a disease of neuronal degeneration like Alzheimer's disease but in a different distribution (the comprehension areas of the temporal lobes at first) and a different pathological category (the fronto-temporal dementias).

The Origin of Speech

Speech of the early representational sort probably developed in the higher reaches of the perceptual representation system and came into being when the representation code reached a certain level of abstraction.

The multi-modal representation areas were the places where this was likely to happen. A multi-modal sensory representation is neither a sight, nor a sound, nor a touch. It is an abstract rendering of a set of primary sensations. In trying to represent or express such a thing, some modes fail: One can neither draw such a thing nor play it on the piano. Speech can do it easily: The phrase, "a soft, mewing, black kitten reeking of fish", describes a multi-modal sensory object perfectly.

The neurologist, Norman Geshwind, first suggested that the multi-modal areas gave rise to the speech system. Bickerton agreed, in that he said that language is only coincidentally a communication system; that it is primarily a representation system, and that, logically, representation must precede communication.

To link multi-modal labels to arbitrary sounds and so create a language map of the sensed world seems a likely step for a vocal animal with an internal representation system. It would allow the coding of complex sensory representations into compact symbolic ones. It would also allow compact memory coding. Imagine the difficulties in storing the linked sensory representations of "a soft, mewing, black kitten reeking of fish".

Did thought representations originate in the multi-modal sensory areas, coded in one of Young's intrinsic brain languages? Did they then, through communication, develop externalized word coding systems going, "See banana. Me eat."? Did they then blunder via Chomsky's propensities into the complex relational structures of syntactic thought and speech?

Speech Sense Organs and Worlds

Verbal like non-verbal concepts can act as Democritian sensory interpretation organs, as well as Lorenzian motor organs, but they increase the scope. They can handle complexities that cannot be dealt with non-verbally.

Words can encapsulate complex concepts and act as what Douglas Hoffstadter calls situation labels. There are fluid situations and sticky situations and win-win and no-win situations. These words are mental models and apprehending in this way is pattern recognition. Problems can arise: Something can turn out not to be a win-win situation.

They can also be thought of as metaphoric labels. A little syntax allows more complex metaphoric labeling. "John is a lion," can mean that the lion is named John, but it can also mean that the man named John is brave. The metaphoric sentence is a different and more complicated kind of animal—to thoroughly mix the metaphor. This metaphor, and this kind of word play, is a long way from, "See banana. Me eat."

Lakoff and Johnson think that we use *metaphoric* thinking much of the time without being aware of it. They use the example sentence: *"I do not have enough time to spend to put these ideas into words."* They point out that without noticing it we have accepted the *metaphors*: *Time is money. Ideas are objects. Words are containers.*

In the above paragraph, all *metaphoric* words and phrases are *italicized* (from Italian script), including the self-referential word, *metaphor* (meta-beyond, phoren-to go).

Space and movement provide metaphors. Our moods and our days and even our lives can go *up and down*—as *the wheel turns.* These entities do not occupy space or move, but we like to interpret reality with such *metaphors.* We also use time *metaphors* and probably will continue to until *time runs out (of the hourglass) and our clocks stop (ticking) at the end of the metaphoric road.* Does the *limbic* space and time percept *framework underlie* this *metaphoric framework*?

We are *metaphoric animals.* We *frame* our thinking with *metaphors,* and the *metaphoric frames* determine how we think. This can, of course, cause problems:

Word labels can *mislead* us: Where does your lap go when you stand up? In an earlier chapter we asked where your fist goes when you open your hand? It takes time to *figure out* that we *frame* the words as objects when they are actions. The label is wrong.

Metaphors can *mislead* us; they can *trip us up.* If you *oppose* this idea, let Lakoff and Johnson *marshal* their facts, *undermine* your intellectual position, and then *assault and demolish* it by using the metaphor: *argument is war*. Let me point out, as they do, how much more pleasant the experience would be with a *metaphoric stance* such as *argument is dancing*.

Metaphors are mental *models* (from the Latin measure) and can function as *speech-concept sense organs* that allow us to recognize *metaphoric patterns*. We are saying again that we use mental *models* of the world to *interpret* the world.

We also use them to *construct* (con-together, struere-heap) our *language* (lingua-tongue) world. The *italicized* words *illuminate* (il-upon, lumen-light) how meanings in the *semantic* (sema-sign) web are built up.

Owen Barfield thought that *concrete* (concretus-condensed) words can be *transformed* (trans-across, forma-mold) over time into *abstract* (ab-from, trahere-to draw away) ones. He noted that the verb, *"to be"* comes from the proto-Indo-European Sanskrit, "to breathe".

The *language* world seems *to be inclined* to *metaphoric* and increasingly *abstract* word *models.*

No more *italics—from here on.*

Gregory's World: The Verbal Concept World

Gregory's world is one of verbal concept models. A new word is a new concept and a new model. As vocabulary grows, the inner model world grows.

Such a verbal concept world is the fourth level of Dennett's generate-and-test hierarchy. He credits Gregory with the idea of verbal mind tools in a mental model map in which Gregorian creatures use words to interpret the world, and "spin complex chains of hypothetical cause and effect before they have to chance action in the real world". This speech world is Craikian, Popperian, and Lorenzian, as well as Gregorian, and can model the real world well enough to test actions in it.

Syntax can be thought of as a mechanism for spinning such chains. It made the concept world better able to elaborate models and predict outcomes. If conditionals did not exist and if we were not able to use subjunctives, would we not be less able to consider hypothetical situations and make predictions? Syntax could be useful enough to drive evolutionary selection.

The non-verbal concept world of the last chapter was a more limited world. It opened new possibilities, but also new pathologies. Those affect the verbal world as well, but there is more potential trouble. There are completely novel speech-world mistakes:

We can make up words that do not mean anything: Zun. Gefliglebock.

We can make up dangerous words: Blasphemer. Outsider. Enemy.

We can make syntactically correct sentences that mean nothing: "Colorless green ideas sleep furiously".

We can make sentences that cannot exist in the world of reason: The circle is square. The man says all men are liars.

We can make verbal and metaphoric mistakes, and quite fundamental ones: My lap goes away when I stand up. Where does it go? My conscious self goes away when I fall asleep. Where do I go?

I am thoughtful. I am articulate. Not when I am asleep. I am neither of those things. I am not really there. I am more like a car sitting in the garage or a computer with the power off. These are not constants but descriptions of the "present state of a variable", according to the philosopher, Willard Van Ormand Quine: a variable state determined by its controlling variables.

We are not objects with fixed properties: Our component molecules are continually being broken down and replaced; our levels of consciousness changed; our mental concepts modified. Heraclitus said that we cannot step into the same river twice: The river has changed, and so have we.

Buckminster Fuller said, "I seem to be a verb". We seem to be, not things, but processes: To be is to breathe. We may have gotten the metaphor wrong.

In the Gregorian verbal world, errors are alarmingly easy to make—and can even be dangerous.

CHAPTER 14

HOMO ABSTRACTION USER AND MEMES AND PLATO'S ABSTRACT WORLD

Two thousand and five hundred years ago the concept of abstraction entered the world. Plato ushered it in. He said there were three separate and distinct kinds of things: physical objects, thoughts and feelings in minds, and abstract things. He called abstract things, Universals or Forms, and he said things about them that muddied the waters, but he did make this grand and fundamental distinction.

Somewhat before Plato, thinkers like Democritus had started to use language for critical thinking and created the abstract philosophical system that Plato worked in.

This was another emergence. The gap between the critical thinking of the Greek philosophers and what came before is so great that there seems to have been an increase in the capacity for thought. It was a coding advance, and a step forward that left other languages behind—languages for horses, as Friedrich the Great once said of German. Latin was such a language: There were no words to discuss philosophy in Latin until Cicero coined them to translate the Greek philosophers.

Children speak a language for horses, simple and concrete. Only at twelve years of age do they become able to deal with abstractions. This is Piaget's fourth stage of cognitive development.

The world of abstractions is not confined to a single nervous system. It is a shared system like speech, common to the entire human race. In our internal representation terminology, it is a shared set of mental models.

Popper's World III

Another philosopher, Karl Popper, also divided the universe into three parts or worlds: World I consisted of physical objects like atoms, or rocks, or people. World II, of the contents of consciousness or mental states such as feelings and thoughts. World III, of the products of the human mind like songs, stories, books, works of art and scientific instruments; and included beliefs, theories of science, and other abstractions.

Popper's worlds intersect and objects can cross categories: Thoughts can be World II and III. Churches and airplanes are part of Worlds I and III, and World II if someone is thinking about them.

World III objects can exist independently of the minds that embodied them: A book written a hundred years ago can be read. They can cease to exist and come back into existence: A Rosetta stone can be incomprehensible for thousands of years, but then comprehended when translated (Popper's example).

World III objects, although products of human thought, can exist—in a sense—even though no mind has ever thought of them or ever will. The product of the human mind known as the number system allows the existence of numbers that are too large to be contained in any human brain, impossible to write down in any finite period, and beyond the digit capacity of any known computer.

The relationship of World II subjective mental states to World III objects is one of conceptual comprehension or grasping of meaning. The mind extracts the meaning just as a perceptual module extracts the meaning of a sensory stimulus. The World III object relates to the World II mind as a World I sensory stimulus does. This is conceptual pattern recognition.

The comprehended World III is learned. It is incomplete. No one mind holds it all. Each personal conceptual world is an Umwelt.

Language and Self

A major World III object is human language (Popper again). Although the brain has predispositions to language, the thing itself is a human mental construction—and a major tool for the development of further constructs. Complex thought is dependent on language: Mathematical thinking cannot be done without its symbol system. Philosophy cannot be done with grunts and gestures. A poem can evoke profound thoughts and feelings, but only through a web of words.

One of these language constructs is the self (Popper yet again). The early self develops before language, but later development occurs through interactions with other selves, and the primary tool of those interactions is language.

The self itself is, to a degree, a World III object. Are Popper and Plato and—more so—Emily Dickinson and Alfred Tennyson not preserved as faded versions of themselves in their writings?

Plato's Abstraction Problem

Plato worried about the status of things like blueness. There is no such thing as blueness in nature; there are just examples of blue things. This led Plato to conceive an abstraction, a universal and perfect blueness called the Form of the Blue. He thought the Form existed in metaphysical space and somehow from there imparted blueness to particular blue objects in the world here below. Blueness was not a product of human thought.

Aristotle disagreed. He thought that the brain took individual examples of blue flowers and derived categories like blue and flower. Blueness, the abstraction, was a product of human thought. It could also be the product of a neural network.

Even if Aristotle was right about Blueness, what was its status before anyone conceived it? Is it possible for an unconceived abstraction to have independent existence? Does Plato's metaphysical world exist?

Popper's Abstract Autonomous Systems

The most abstract of World III abstractions are mathematical objects and with them Plato comes closest to making his argument for metaphysical Forms. The idea of the perfect Form of a horse is odd, but the perfect Form of the triangle with all real triangles just crude approximations seems plausible. Would triangles exist if our minds did not contain the concept? Would there just be meaningless arrangements of lines?

What about mathematical truths? Do we invent them, or do they exist independently, and we only discover them? What about the number zero? We did math for thousands of years without it. It was first used by a Hindu mathematician around 700 A.D.. Did it exist before the mathematician thought of it?

What about the square root of minus one? It has no conceivable physical meaning and yet it exists. It not only exists; its square has a physical meaning.

Popper's answer is that World III is to a degree autonomous. We create a thing like the number system, but then it takes on a life of its own. It has an objective reality and objective properties that exist latently within the system even if we do not realize they are there. I think Popper would say that a triangle has independent and autonomous existence, but only after we have created the abstract autonomous system of geometry; until then, it is just a meaningless set of lines.

Popper did not believe that Plato's metaphysical world existed. He thought that concepts prior to experience did not exist. He thought that concepts of this complexity could only come into existence only after we had developed an abstract thought system, a mental model system, that could allow them.

This suggests that not only triangles, but philosophy and neurology, good and evil, morality and religion, and even the gods themselves come into existence only after a mind has created a concept system that allows them. Do dogs have gods? What about tunicates? If triangles had gods, would they have three sides?

This also suggests that our seminal creative act is the making of an abstract autonomous model of some aspect of the outer world. Once manufactured, the system can be explored, and its full capabilities and autonomous properties—and its meanings—discovered and exploited.

Popper suggests but does not prove that Plato's metaphysical world does not exist. There could still be Forms and Gods. They could have existed before our concept systems could conceive such abstractions and continue to do so. A physical entity cannot prove a metaphysical entity does not exist. It cannot prove one does either.

Mathematical Abstract Autonomous Systems

A mathematical system is the clearest type of abstract auto nomous system. It is a rule driven language with objects like straight lines or numbers, axioms like parallel lines cannot meet, and syntactical operations. The axioms and operations determine every possible consequence: If x is true, then y can be proven to be true or false—and y can be an unrealized, autonomous property. If the number one and the subtraction operation are given, then (1-1) can be written and zero exists, even if we do not realize it for thousands of years. If the square root operation is given, then the square root of minus one exists.

These are limited, closed systems. They are called formal systems. There is nothing novel in such systems once you have decided on the axioms and operations, there is only playing around and finding all the results.

A computer program can act as an autonomous abstract formal production system. It can grind out all possible mathematical statements allowed by a set of axioms. The computer would not know what a mathematical object was or what its output results meant or if they meant anything.

A mathematician, like a computer, can do mathematics as a set of abstract syntactic operations with no relevance to the outside world, but the mathematician can also interpret them to have meaning. The World II of a human interpreter can use World III geometric concepts like triangles to construct World I houses.

Automatous consequences matter. If parallel lines do not meet in Euclidian geometry, then the mathematician cannot take that system and use it to navigate on the surface of a sphere where parallel lines do meet. The mathematician must jump out of the Euclidian formal system and invent spherical geometry.

It is difficult now to appreciate the effect that mathematical thinking has had on all of us, even those of us who cannot do it. The first instrument for investigating reality was theology, but even theology was influenced by mathematics. Xenophanes around 500 BC dispensed with multiple Gods by arguing that there could only be one God because God was perfect and so secondary Gods would have to be imperfect and were thus impossible. He dispensed with the anthropomorphic God by arguing that the one God would be "...the same from all sides...otherwise the various parts would be superior and inferior to each other, and this is impossible…", and thus the one God would be a sphere.

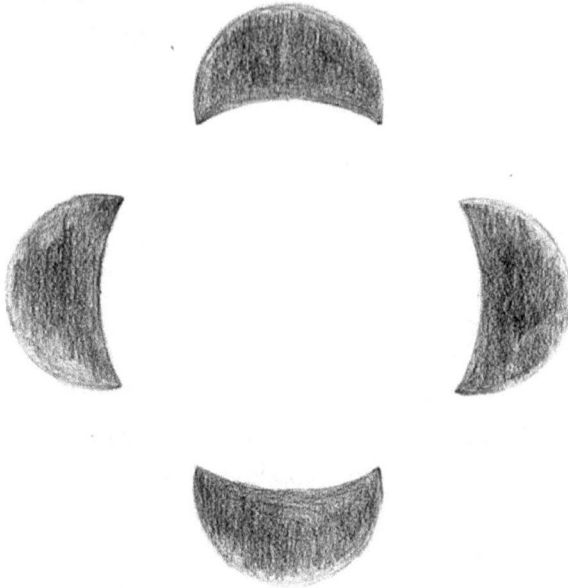

Figure 140 Xenophanes' Hypothesis

Xenophanes might have made the same argument for the universe, or even the Self.

In the Middle Ages, the anthropomorphic God was still around, and the instrument of investigation was still the abstract autonomous system of logical theology—still influenced by mathematics: God would logically place the earth at the center of the universe. God would logically move the planets around it in perfect circles. God would logically place perfect stars up above in a perfect heavenly sphere.

The scientific revolution changed all that. Copernicus, in 1543, moved the earth from the center of the universe. Kepler found orbits to be ellipses. Galileo observed the heavens to be imperfect. The focus shifted from God's-purpose-was-this teleology to purpose-free and largely mathematical descriptions. The only criterion for those descriptions was that they could model the outer world well enough to allow predictions. They were a form of predictive coding.

Mathematical science became the new abstract autonomous system. Descartes, with the invention of analytic geometry and the x-y plot, made space an abstraction that could be described numerically. Newton and Leibnetz, with the invention of calculus, made movement through space an abstraction that could be described numerically. Newton and Laplace made the movement of planets and suns through space a mathematical abstraction that could be described numerically. When Napoleon asked about God, LaPlace replied, "I have no need of that hypothesis, sire."

Earlier, we took a verbal statement about unique receptor responses allowing recovery of unique memories and coded it into the mathematical representation

$$fm = (\Delta A' + \Delta A''). r'$$

$$= (f'. r' + f''. r''). r'$$

$$= f'. r'. r' + f''. r''. r'$$

$$= f'.$$

We were able to make the movement of a memory in and out of memory space into a mathematical abstraction that could be described numerically.

We were then able to see that the requirements for an r'-input to retrieve its f'-memory from the content addressable memory was that the r'. r'-operation was equal to one.

We were able to "see": The mathematical statement opened a window into the abstract problem. The solution presented itself.

A mathematical statement can be an abstract sensory organ. If the window of the sensory organ is clear, we can see into the depths of the problem.

A mathematical statement can be an abstract model. If the model's correspondence with reality is accurate, it becomes "a working physical model of reality" with a "relation-structure" to the outer world which "works in the same way as the process it parallels", and the syntactic operations roll out flawlessly.

Our notation allowed us to make a working model of content addressable memory. Newton's three laws allowed him to make a working clockwork model of the universe.

A mathematical model is a Craikian mental model that can allow Gregorian hypotheses about inputs and outputs. It can allow predictive coding and anticipation of future events.

Hegel's perfect apprehension may require perfect description in perfect mathematical language. Perhaps some future synesthete will be able to directly apprehend it. We may all have to learn mathematics.

The Natural Thought and Language System and Analogy

What about the natural thought and speech system we use every day to deal with the world? It is not a formal system. It is sloppier with poorly-defined terms and gaps and inconsistencies.

It is, however, an abstract system with similarities to a mathematical one: It has objects called nouns and operations called verbs. It has syntactical rules that allow lawful constructions called sentences. It has subjunctive moods and conditional if-then sentences for logical thinking and prediction. Usually though, we do not bother with logic and just use pattern recognition.

It has more flaws than a formal mathematical system. It has built-in problems:

Meaningless words are not a problem in mathematics but certainly are in language: Phlogiston was a substance within physical objects that caused them to burst into flame until Lavoisier discovered oxygen—not in the objects themselves, but in the air around them.

Meaningless sentences are easy to produce, and a problem: Think of jokes about politicians.

Impossible and self-contradictory sentences are a problem: This sentence is wrong. I am a liar.

Labels and metaphors are a problem: Something can turn out not to be a win-win situation.

Automatous system properties are a problem: If discussion is war, then bad things can happen at faculty meetings.

Such systems are like perceptual modules, they allow but also limit conceptual rather than perceptual possibilities.

One recurring theme is that every advance in cerebral processing comes at the price of new mistakes. Abstract thinking is no different.

Curiously, not all of natural thought and language mistakes are mistakes. We use self-contradictory sentences all the time, as I just did, and they mean something. I am not myself. I am beside myself. We are one. These statements are self-contradictory and irrational—but you understand them.

We use metaphors all the time and they can be irrational and sometimes self-contradictory: Moods do not go up and down. John is not a real lion. But you understand.

Self-contradiction and irrationality carry meaning. Formal system language carries meaning, but only denoted or defined meaning. Natural language has capabilities beyond the formal. Its words and metaphors can "carry beyond" formal meaning. They connote as well.

Natural language employs the semantic web. It pulls meanings from strands of James's "huge spider's web of silken threads suspended in the chambers of consciousness". This is where we pull the strands together and say this is like that. This is where we make metaphors and analogies.

Steven Hofstadter argues that we think largely by analogy. In the last chapter we touched on his situation labels. He has made a broader case for analogy as the core of thinking. He thinks that we frame experience, not only with word labels, but with proverb labels like "a stitch in time saves nine", personal history labels like "the time Andy tried to catch the cat", and literary labels like a Pyrrhic victory.

He argues that our basic conceptual operation is this-is-like-that, and that that "every concept we have is essentially nothing but a tightly packaged bundle of analogies, and...all we do when we think is...leap from one analogy-bundle to another". In the land of analogy, we are sometimes beside ourselves, as Andy was when he tried to catch the cat, and not every mistake is a mistake, although that one was.

He argues that, as we age, we build up our analogy bundles to allow us to perceive more complex and more abstract patterns of this-is-like-that in the world, and "chunk" the world into bigger and more abstract patterns: Andy did eventually manage to catch the cat, but it was a Pyrrhic victory.

We have argued that the brain does pattern recognition: Perception is pattern recognition. Barlow's primary operation of the cortex is pattern recognition. Hofstadter's contention is that thinking is the recognition of analogous patterns.

We could say that the semantic web is composed of mental models, and that we recognize analogous patterns in the outer world.

We could say that the chunking operation tends to move us from the episodic to the semantic. We could worry that this leaches the specific and vivid instants from our memories and banks the fires of youth.

We could say that the web is not a formal but a human system and part of what we are. It is where we make mental worlds and where they change and age along with us.

The Processing Perspective

We could say that the web contains mental models that allow pattern recognition but also larger contexts that frame our thinking and pattern recognition.

We discussed the effect of limbic contexts. Abstract thoughts have abstract contexts. The sensory priming effect of a sea story is to see "hal" and think "halibut". Such contexts can be thought of as frameworks or larger mental model worlds that frame our mental models. They can assist, but also distort, thought. Xenophanes and Descartes have provided examples.

These processing perspective sections have been a deliberate attempt to impose a signals processing context on nervous system function, a framework outside the biological one. A deceptive aspect of the imposed framework has been the two-dimensional one-sided sketch of the nervous system. The sketch below is to remind us that it is three-dimensional and has side-to-side and sensory-motor splits.

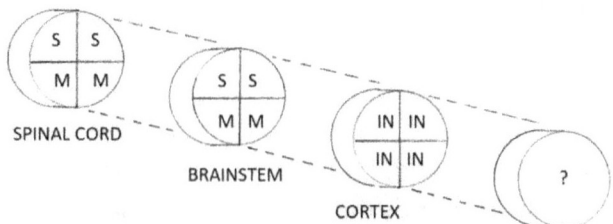

Fig 141 The bilateral and superior-inferior sensory-motor splits of the central nervous system with a reminder that the brain is also split and contains only interneurons.

It is also drawn to beg a question: We do not experience the world in two halves, so where is the unified experience? Where is the processing space where the me-self and the world model interact as single entities?

The sketch suggests it is in the frontal lobes, but we have found no such frontal area. Is it undiscovered? Might Descartes be right about the pineal gland? Is it a bad question imposed by the sketch framework? Is the unified-space really a functional space, a distributed set of interconnected areas that function as one? Is our basic framework for interpreting the sensed world a perceptual distortion—a mental model illusion?

What about the mental space reported by mystics where we are not only one, but one with the universe? Dostoevsky experienced such a space during his epileptic seizures. A trained observer, the neuroanatomist, Jill Bolte Taylor, had a cerebral bleed with intermittent compression of her left hemisphere. She described flipping back and forth between clear organized thought and self-apartness, and total absence of speech and internal chatter and complete one-ness with all. She felt enlarged and free and like "...a great whale gliding through a silent sea of euphoria." Nirvana seems to be in the right hemisphere and available to anyone who can find a cooperative neurosurgeon.

The evidence of our sensory systems is indirect and imperfect. To evaluate it reliably requires conceptual understanding of those systems. We must be skeptical of perception, mindful of context, careful with analogy, and critical in thinking.

We must be careful with our models. To deal with the complexity of the nervous system, I chose to code it into an abstract signal processing model to allow me to see through the "blooming buzzing confusion" and extract meaningful patterns from the noise. The mental model hypothesis says that the nervous system, faced with the complexity of the real world, does something similar.

My model led me into the unified-space problem a few paragraphs back. Did my model distort my thinking? Would another model have done better?

A statistician, George E. P. Box, is said to have said, "All models are wrong, but some are useful". As to this one's usefulness, the reader is now in a position to decide.

Mental Models and Memes

Thoughts travel.

Once men had minds, they had mental objects in them called thoughts. Once they had language, they were able to transmit those mental objects to other minds. These transmittable mental models could then take on lives of their own. They could multiply in many minds. They could live in new minds after the original thinking mind had died. They could change. They could be forgotten and cease to exist.

Richard Dawkins called them memes, which he defined as units of cultural transmission. Memes fly about the world and lodge in peoples' heads. If someone points and says dog, you have a new representation concept. If someone hoists with a block and tackle, you have a new motor concept. If someone says unconscious mind, you have a changed concept of yourself and your Self. With memes, you can interpret the world, not only in the light of your own experience, but of others as well.

Dawkins' theory has attracted a lot of attention and elaboration and some controversy about what memes are and can do and even whether they exist. We are going to concern ourselves only with the uncontroversial concept of a transmittable mental model or idea.

Transmittable Ideas and Idea Systems and Superorganisms

Transmissible ideas had been discussed before Dawkins. T.H. Huxley in 1880 proposed that they could undergo something like Darwinian evolution as they bounced around from person to person. This was probably how abstract thought came into the world: The brain did not change (in any way we can detect). The ideas evolved.

Ideas can be transmitted without language: One can learn by watching. Stone tool chipping was transmitted this way for a million years. This was limited and slow and had little effect. Speech speeded the process up and cultural evolution accelerated past biological evolution.

Ideas can come as systems of related items. We live in a world largely constructed of such systems. Although farmers work with World I dirt, their farming methods are ideas from World III. Although our houses and offices are made of World I bricks, their functions are World III ideas.

A set of ideas constitutes a job like carpentry or a scientific discipline like geometry. A set of ideas constitutes a social system like a bridge club or a religion.

A set of ideas allows a group of individuals to work as a collective entity, a superorganism. We are not the only species able to do this. Ants and bees can form superorganisms, but theirs are fixed and inflexible. Only we can do so at will, and with concepts.

A modern corporation is a set of ideas. It is, literally, the embodiment—incorporation—of an idea for making money by selling cars or computer searches. It is a superorganism, and the work of its employees can be the manipulation of its ideas on idea-constructs called computers.

A society is a set of ideas. Its evolution is the evolution of its ideas and what it leaves behind when it fades away. As Mary Elizabeth Coleridge put it almost two hundred years ago,

> ...Glorious Rome has lost her crown
> Venice' pride is naught
> But the dreams their children dreamed
> Fleeting, insubstantial, vain
> Shadowy as the shadows seemed
> Airy nothings, as they deemed.
> These remain.

As hers has.

Mental Model Systems and Their Discontents

A cultural set of ideas is a shared mental model. Its societal superorganism is the realization of that model, and its motor organ.

The model reflects biology and history, and its concepts should reflect relevant threats and rewards. A tribal society has war gods and fertility goddesses, soldiers and priestesses, hunters and gatherers,. We have farmers, doctors, armies, and pension plans.

Why do we have flat-earthers and alien-abduction-believers?

As we go up the ladder of sensory processing, the direct feedback of the outside world diminishes, and the possibility of error or irrelevance increases. A receptor can only fail to register a stimulus. A perceptual system can manufacture a false one like a phantom limb. A conceptual system can manufacture frank delusions. Ideas can grip us to the point of overwhelming the basic biological motivations. We can decide to die for an intellectual abstraction such as a religion or a country.

As a cultural model becomes organized around World III, the threats and rewards of World I become less relevant. The feedback that keeps the model accurately reflecting the outside world diminishes. What is the significance of the weather to an urban office worker? The inhabitants of the cultural model can become more influenced by the model than the outside world. Fashion models can induce teen-age eating problems.

A cultural mental model is a Popperian abstract autonomous system. It can have a life of its own. It can have autonomous properties with consequences.

It can distort apprehension of the outer world. Theology can require that the earth be at the center of the universe, dissenters be executed, and religious wars be fought. Marxism can re-define capitalism as class oppression, and class warfare to seize the means of production be justified.

It can kill. The realized concept of a motor vehicle has killed millions of us. The utopian fantasy of Marxist socialism killed eighty million people in Russia and China in the last century. The toll of religious wars is something best not to think about.

Only we can construct concept systems. Only we can construct model systems that go wrong and superorganisms that become dysfunctional. How do we prevent this? How do we keep conceptual and cultural evolution on track?

Thinking clearly seems to be the answer. Scientific rationalism, the program. The scientific method, the mechanism. Looking for error and testing theories for validity—or Popper's falsifiability—seems to be the only way of keeping conceptual models grounded. The success of western civilization since the Enlightenment seems to validate this.

The Scientific Model Universe and Plato and Popper's World

A biological entity is a model of the world it inhabits. J. Z. Young said: "A wing reflects the existence of air; a fin, water...The whole organism can be considered a coded representation of its environment."

The brain is a biological entity, and that it is a model of the world is the major theme of this book.

The World III of products of human thought is a biological entity, although a somewhat odd one, and a model of the world it inhabits—Plato and Popper's world. It is universal: It can represent anything and everything. Although disembodied and shared, it is the eighth functional part of every human nervous system.

Its scientific model is an evolving project devoted to representing, not only the world, but the universe. Its scientific method is the fifth and final level of Dennett's generate-and-test hierarchy: a formalized way of "generating theories and testing them in public so others can inspect them and making mistakes in public so others can avoid them".

This is not to say this model is perfect, but it is perfectible—or at least near-perfectible. There may or may not be progress in history, but there is progress in the model. It is our evolving map of the universe, our release from Plato's cave, our doorway into Hegel's perfect apprehension and execution.

We move slowly toward better apprehension and representation. When I was an intern, I went to a lecture on Zen Buddhism and encountered a meme that became part of this chapter. The lecturer, whose name I have long since forgotten, concluded with the marvelous phrase, "We are the organ by which the universe regards itself."

I think she was right. I think we are getting better at it.

REFERENCES AND NOTES

Chapter 1. Nervous Systems and Models

...100,000,000,000 neurons: No one has counted them. There are various estimates. This is the one quoted most often. There are various others depending on whether only the cerebral hemispheres are counted or whether the cerebellar hemispheres and the brainstem are included. Then there are the billions of gut-related neurons that have been called a gastrointestinal brain. Then there are the billions of glial support cells.

...ontogeny recapitulates phylogeny: Ernst Haeckel in 1892 said," Ontogeny is the brief and rapid recapitulation of phylogeny." Quoted in Gregory, R.L., "Editorial: neuroarcheology", Perception 29, 505-508, 2000.

...embryo drawings: After the originals in Haekel, E., Neterlichte Schapfurgeschichte, Reiner, Berlin, 1868.

...neuron depictions: The drawing of the pyramidal cell is after one in pen and ink by Santiago Ramon y Cajal in "The Croomian Lectures: La fine structure des centres nerveux", Proc. Roy. Soc. Lond. Ser. B, 55, 444-467, 1894.

...enchanted loom...synapse: Pearce, J.M.S., "Sir Charles Sherrington (1857-1952) and the synapse", J. Neurol. Neurosurg. Psych. 75, 544, 2004. The original reference is Sherrington, C., The Integrative Action of the Nervous System, Scribners, 1906.

...numbered brain regions: This drawing is in every neuroanatomy textbook. The original is in Brodman, K., Localization in the Cerebral Cortex, Leipzig, 1909.

...action potential: A drawing like this is in every neurophysiology textbook. See Hodgkin, A.L. and Huxley, A.F., "A quantitative description of membrane current and its application to conduction and excitation in nerve". The Journal of Physiology. 117 (4), 500–44, 1952, and Hodgkin, A.L., The Conduction of the Nervous Impulse, Thomas, Springfield, IL, 1964.

...model neurons: McCullough, W. and Pitts, W., "A logical calculus of the ideas immanent in nervous system activity", Bull. Math. Biophysics, Vol. 5, Univ. Chicago Press, 1943.

...summation model neuron of the 1970's: Anderson, J.A., "Two models for memory organization using interactive traces", Math. Bioscience 8, 137-160, 1970, and Anderson, J.A., "A simple neural network generating an interactive memory," Math. Bioscience 14, 197, 1972; and Cooper, L.N.," A possible organization of animal learning and memory", in Lundquist, B. and Lundquist, A. (eds), Proceedings of the Nobel Symposium on the Collective Properties of Physical Systems, Acad. Press, NY, 1972, p. 252-264; and Kohonen, T., Associative Memory: A System-Theoretical Approach, Springer-Verlag, Berlin Heidelberg New York, 1978.

...auto-associative neural network: Kohonen, T., Self-Organization and Associative Memory, Springer Verlag, Berlin, 1984.

...self-organizing neural network: Kohonen, T., "Self-organized formation of topologically correct feature maps", Biol. Cybernetics 43, 59-69, 1981.

"...in common with single cellular organisms...": Livingston, R.B., "Sensory processing, perception and behavior", in Grenell, R. and Gabay, S. (Eds.), The Biological Foundations of Psychiatry, Raven Press, NY, 1976.

...receptor: The concept was suggested by J. Langley in 1898. His work and the work of P. Ehrlich is discussed in Rang, H.P., "The receptor: Pharmacology's big idea", Br. J. Pharmacology 147(Suppl 1), S9-S16, 2006.

...reflex: Rene Descartes first used the concept and term in a book called De Homine in 1662.

... freedom from tyranny of reflex: Gregory, R.L., "Knowledge in perception and illusion", Proc. Phil. Trans. R. Soc. Lond. B 352, 1121-1128, 1997. Gregory also said that the time prediction allowed by such a perceptual system "frees the animal from the tyranny of control by the reflexes, to allow intelligent behaviour into anticipated futures", in Gregory, R.L., "Perceptions as hypotheses", Phil. Trans. R. Soc. Lond. B 290,181-197, 1980.

...only physical correlate of biological intelligence: Karl Lashley said:" The only neurological character for which a correlation with behavioural capacity in different animals is supported by significant evidence is the mass of tissue, or rather the index of cephalization...which seems to represent the amount of brain tissue in excess of that required for transmitting impulses to and from the integrative centers", in Lashley, K., "Persistent problems in the evolution of mind", Quart. Rev. Biol. 24, 28-42, 1949.

...brain weight versus body weight: Jerison, H.L., "Animal intelligence as encephalization", Phil. Trans. Royal Soc. Lond. B308, 21-35, 1985.

...variation in brain size: Jerison, H.L., "Brain evolution: new light on old principles", Science 170, 1224-1225, 1978.

...human brain size: Jerison listed the average weight is 1350 grams. The adult male is said to average 1370 grams and the female 1200 grams, and there are variations such as the larger visual cortices of northern populations.

...saw the neuron that did the work of the brain: Cajal, S.R., "The Croomian Lectures: La fine structure des centres nerveux", Proc. Roy. Soc. Lond. Ser. B, 55, 444-467, 1894. (They had been seen before, but Cajal's neuron stains and images were good enough that he could say that neurons were probably separate cells—and describe them as "the butterflies of the brain". The alternative hypothesis was that they were all connected and formed a continuous network or reticulum, and this was conclusively proved wrong only with the electron microscope pictures of synaptic cleft gaps in the 1950's.)

Chapter 2. Pre-vertebrates and Stimulus-Response and Receptor World

...embryonic tunicate: Manner, H.W., Elements of Comparative Vertebrate Embryology, McMillan, NY, 1964, and Dimond, S.J. and Blizard, D.A., Evolution and Lateralization of the Brain, Ann. New York Acad. Sci. ANYAA9, 299, 397-410, 1977.

...a primitive worm flipped itself over onto its back, found that it worked better, and stayed that way: Huxley, T.H., Lectures on the

Elements of Comparative Anatomy, Churchill and Sons, London, 1864. Quoted by but thought simplistic by, Jolie, M., "The origin of the vertebrate brain", Ann. New York Acad. Sciences 299,74-86, 1977.

...an eyespot and statocyst: Young, J.Z., The Life of the Vertebrate, Clarendon Press, Oxford, 1981. (The nervous system reabsorption drawing is after one in this book.)

...lancelet: Manner, H.W., Elements of Comparative Vertebrate Embryology, McMillan, NY, 1964, "Chapter 7: The development of amphioxus".

...gene segments...vertebrate pattern is latent in this primitive precursor: Holland, L.Z. et al," The amphioxus gene illuminates vertebrate origins and cephalochordate biology", Genome Res. 18(7), 1100-1111, 2008. The gene duplication idea originated in Ohno, S., Evolution by Gene Duplication, Springer, Berlin, 1970.

...Umwelt: Von Uexkuell, J., Theoretical Biology, Harcourt and Brace, New York, 1926. Quoted in Allman, J.M., Evolving Brains, Freeman and Co., NY, 1981.

..." long way from amphioxus": This has been called The Fighting Song of The University of Ediacara, after the Ediacarian geological epoch, and attributed to various people including a professor at Texas A and M, named Sewell H. Hopkins, around 1921.

Chapter 3. Jawless Fish and Smell Brain and Darwin's World

..." And then a miracle occurs." Sidney Harris cartoon caption. In Mankoff, B., Complete Cartoons of The New Yorker, Black Dog and Leventhal, New York, 2004.

...head placodes...development of the adult brain: Butler, A.B. and Hodos, W., Comparative Vertebrate Neuroanatomy, Wiley and Sons, New Jersey, 1996.

...olfactory bulb...smell receptors: Davis, J.L. and Eichenbaum, H., Olfaction: A Model System for Computational Neuroscience, MIT Press, Cambridge, Mass., 1991.

...olfactory bulb and cortex: Shepherd, G.M., The Synaptic Organization of the Brain, Second Edition, Oxford Univ. Press, 1979.

...such pathways from the olfactory system to the brainstem motor pursuit system: Derjean, D. et al, "A novel neural substrate for the transformation of olfactory inputs to motor outputs", PLoS Biol. 8(12): E1000567, doi:10.1371 /journal.pbio, 1000567, PLIOS.org, 12/21/2010, and Kermen, F., Franco, L.M., and Yaks, E., "Neural circuits mediating olfactory-driven behavior in fish", Frontiers in Neural Circuits 7, 62, 2013.

...reticular formation: Moruzzi, G. and Magoun, H.W., "Brain stem reticular formation and activation of the EEG", Electroencephalography and Clinical Neurophysiology 1: 455-73, 1949.

...replicating toolkits: Dobson, V.G., "Models and metaphysics: the nature of explanation revisited". In Rose, D. and Dobson, V.G. (eds), Models of the Visual Cortex, J. Wiley and Sons Ltd, 1985.

Chapter 4. Jawed Fish and Pattern Recognition

...family of receptor gene copies: Buck, L. and Axel, W., "A novel multi-gene family may encode odor recognition: the molecular basis for odor recognition", Cell 65,175-187, 1991.

...cortex...modules...: Mountcastle, V.B., "The neural mechanisms of cognitive functions can now be studied directly", Trends in Neuroscience, 505-507, Oct. 1986. (His earliest paper on this subject was in the 1950's.)

...olfactory bulb...oscillates: Shepherd, G.M., "Computational structure of the olfactory system". In Davis, J.L. and Eichenbaum, H. (eds.), Olfaction: A Model System for Computational Neuroscience, M.I.T. Press, Cambridge, Mass., 1991.

...early theory of the smell system...a pattern: Adrian, E.D., "Sensory discrimination with some recent evidence from the olfactory organ", Br. Med. Bull. 6, 330-331, 1950.

...oscillations...other sensory...memory areas: The oscillation binding theory was first described in the visual cortex in Gray, C.M. and Singer,

W.., "Stimulus-specific neuronal oscillations in orientation columns of cat visual cortex", Proc. Nat. Acad. Sci. USA 86:1698–1702, 1989.

...topographic display...general representations: Lynch, G and Baudry, M., "Synapses, Circuits, and the Beginning of Memory, MIT Press, Cambridge, 1986.

...vector coding: Pellionisz, A. and Llinas, R., "Brain modeling by tensor network theory and computer simulation. The cerebellum: Distributed processor for predictive coordination", Neuroscience 4,323-348, 1979. (See also the pattern representation discussion in Churchland, P., Neurophilosophy, MIT Press, Boston, 1986.)

...pattern recognition: Fisher, R.,, "Linear discriminant analysis", Ann. Eugenics 7(2).179-188, 1936.

...experimental olfactory pattern recognition: Gelman, R.L. in Menini, A.(ed), The Neurobiology of Olfaction, CRC Press, Boca Raton, Fl, 2010.

Chapter 5. Fish and Learning Patterns and Connectionism and Hebb's World

...smell...spots...olfactory bulb: Adrian, E,.D., "Sensory discrimination with some recent evidence from the olfactory organ", Br. Med. Bull. 6, 330-331, 1950.

...Freeman experiments: Freeman, W.J. and Schneider, W., "Changes in spatial patterns of rabbit olfactory EEG with conditioning to odors", Psychophysiology 19, 44-56, 1982.

...tuning effect...in real jawed fish: Atema, J., and Derby, C., " Ethological evidence for search images in predatory behavior". In Grastyan, E. and Molnar, P. (eds), Adv. Physiol. Sci., Vol.16, Sensory Functions, 1980.

...sensory priming: Budson, A.E. and Price, B.H., "Memory: clinical disorders", Encyclopedia of the Life Sciences, Macmillan Publishers Ltd., Nature Publishing Group, 2001.

...Hebb neural network learning rule: Hebb, D., The Organization of Behaviour, J Wiley and Sons, NY, 1949.

...recurrent loops...input-output interaction: Di Prisco, G.V., "Hebb synaptic plasticity," Progress in Neurobiology 22, 80-102. 1984.

...The Network is now a Model of the Outer world: This idea follows from the concept of the organism being a coded model of its environment in Young, J.Z., Programs of the Brain, Oxford University Press, Oxford, 1978.

...distributed representations: Kohonen, T. and Plate, T., "Distributed representations", Encyclopedia of Cognitive Science, Macmillan Reference Ltd, 2003.

...connectionist distributed memories: Kohonen, T., Associative Memory: A System-Theoretical Approach, Springer-Verlag, Berlin Heidelberg New York, 1978.

...distributed memory damage tolerance: Anderson, J.A,., "Neural models with cognitive implications". In LaBerge, D. and Samuels, S.J. (eds), Basic Processing in Reading, Perception and Comprehension, Erlbaum, Hillesdale, N.J.,1977, pp. 27-89.

...input damage tolerance: Kohonen, T. and Lehtio, P. et al, Neuroscience 2, 1065, 1977.

...incomplete inputs and completion: Kohonen, T., Self Organization and Associative Memory, 2nd Edition, Springer-Verlag, Berlin Heidelberg New York, 1984.

...noise tolerance: Anderson, J.A., "Neural models with cognitive implications". In LaBerge, D, and Samuels, S.J. (eds), Basic Processing in Reading, Perception and Comprehension, Erlbaum Press, Hillsdale, N.J,1977.

..." clouds coming down over the mountains": A frequent quote on neurology rounds.

...survive the learning experience: Popper, K., The Self and Its Brain, Springer, Berlin, 1977.

Chapter 6. Amphibians and Sensory Processing Transforms and The Brain as a Cerebral Coding and Language Problem

...coelacanth: Thomson, K.S., Living Fossil: The Story of the Coelacanth, Norton and Co., New York, 1991.

...frog's eye tells...brain: Levin, J.Y., Maturana, H.R., McCullock, W.S., Pitts, W.H., "What the frog's eye tells the frog's brain", Proc. of the IRE 43 (11), 1940-1959, 1959.

...squashed-flat, one-eyed frog: This example is modified from one in Churchland P., Neurophilosophy, MIT Press, Boston, 1986.

...brain as a coding and language problem: Young, J.Z., Programs of the Brain, Oxford Univ. Press, 1978. Also see, Fodor, J., The Language of Thought, Harvard University Press, 1975.

...algebraic brain model...puts the ΔA's individual ΔA's in the right positions: Jordan, M.I., "An introduction to linear algebra in parallel distributed processing", in Rumelhart, D.E. and McClelland, J.L., Parallel Distributed Processing, Volume 1, MIT Press, Cambridge, Mass., 1987.

Chapter 7. Early-reptiles and Biological Content Addressable Memory

...early-reptiles: McLoughlin, J., Synapsida, Viking Press, NY, 1980.

...signature and threat display: Carpenter, C.C., "Ritualistic social behaviors in lizards". In Greenberg, N. and McLean, P.D. (eds), Behaviour and Neurology of Lizards, NIMH, 1978.

...brain endocast drawing: After Witmer, L.M. and Ridley, R.C., "New thoughts on the brain, braincase, and ear regions of tyrannosaurus rex...", Anatomical Record 252 (9), 1266-96, 2009.

...hippocampus...reptiles: It is called the medial pallidum in reptiles and that it functions like the mammalian hippocampus is discussed in Butler, A.B. and Hodos, W., Comparative Vertebrate Neuroanatomy, Wiley and Sons, New Jersey, 1996.

... a map with synaptic memory: Kandel, E.R., "The biology of memory: a forty-year perspective", Journal of Neuroscience 29(41), 12748-12756, 2009. (The best-studied hippocampal form of memory is called long term potentiation. See Lynch, G. and Baudry, M., "The biochemistry of memory: a new and specific hypothesis", Science 224. 1057-1061, 1984.)

...place cells...map: O'Keefe, J. and Nadel, L., The Hippocampus as A Cognitive Map, Clarendon Press, Oxford, 1978.

...size of the hippocampus varies considerably among species: Sherry, D.F., Jacobs, L.F., and Gaulin, S.J.C., "Spatial memory and adaptive specialization of the hippocampus", Trends in Neuroscience 15, 298-302, 1992.

...egocentric map...animal localization cells: Hartley, T., Lever, C., Burgess, N. and O'Keefe, J., "Space in the brain: how the hippocampal formation supports spatial cognition", Philos. Trans. R. Soc. Lond. B Biol. Sci. 369(1635), 2014.

...hippocampus size...London cab drivers: Macguire, E.A. et al, "Navigation-related structural change in the hippocampi of taxi drivers", Proc. Natl. Acad. Sci. USA 97(8), 4398-4403, 2000.

...human spatial map problems: Iario, G,. et al, "Developmental topographic disorientation: case 1", Neuropsychologia 47(1), 30-40, 2009.

...content-addressable memory: Kohonen, T., Associative Memory: A System-Theoretical Approach, Springer-Verlag, Berlin Heidelberg New York, 1978.

... connectionist content addressable memory recovery model...tricky r-vector property: For those who remember linear algebra and prefer standard notation, vectors are column vectors, and vector multiplication transposes the first vector so that r1 multiplies r'1 and r2 multiplies r'2 and the two are added to result in a scalar numerical value

$$r. \, r' = r^T. \, r' = [r1 \; r2]. \, [r'1] = r1.r'1 + r2.r'2$$
$$[r'2]$$

so

$$r. r = r^T. r = [1\ 0]. \begin{bmatrix} 1 \\ 0 \end{bmatrix} = 1$$

and

$$r'. r' = r'^T. r' = [0\ 1]. \begin{bmatrix} 0 \\ 1 \end{bmatrix} = 1$$

and

$$r. r' = r^T. r' = [1\ 0]. \begin{bmatrix} 0 \\ 1 \end{bmatrix} = 0$$

and the ΔA matrix is made with a transpose of the second vector and multiplication then produces a matrix,

$$\Delta A = f. r^T = \begin{bmatrix} f1 \\ f2 \end{bmatrix}. [r1\ r2] = \begin{bmatrix} f1.r1 & f1.r2 \\ f2.r1 & f2.r2 \end{bmatrix}$$

and so

$$f = A''''. r' = \Delta A'. r' = f'. r'^T. r' = f'.$$

See Jordan, M.I., "An introduction to linear algebra in parallel distributed processing", in Rumelhart, D.E. and McClelland, J.L. (ed's), Parallel Distributed Processing, Volume 1, MIT Press, Cambridge, Mass., 1987.

...H.M. had both hippocampal regions removed: Scoville, W.B. and Milner, B., "Loss of recent memory after bilateral hippocampal lesions", J. Neurol. Neurosurg. Psychiatry 20, 11-21, 1957.

...Kluver-Bucy syndrome: Kluver, H. and Bucy, P.C., "'Psychic blindness' and other symptoms following bilateral temporal lobectomy in rhesus monkeys", Am J Physiol. 119, 352-353, 1937.

...context memory: Maren, S., Phan, K.L., Liberzon, I., "The contextual brain...". Nature Reviews Neuroscience 14(6), 417-423, 2013.

...in Alzheimer's disease: Kile, S.J,. et al, "Alzheimer's abnormalities of the amygdala with Kluver Bucy Syndrome symptoms," Arch. Neurology 66(1), 125-129, 2009.

...hippocampal representations...paleo-brain: Lynch, G. and Baudry,

M., "Synapses, Circuits, and the Beginning of Memory, MIT Press, Cambridge, 1986. Discussed in Wilson, D.A. and Rennaker, R.L., Chapter 14 of Menini, A. (ed), The Neurobiology of Olfaction, CRC Press, Boca Raton, 2010.

...qualification...hippocampal representation ...episodic memory: Kandel, E.R., "The biology of memory: a forty-year perspective", Journal of Neuroscience 29(4), 12748-12756, 2009.

..hydraulic psychological theory: Freud, S., An Outline of Psychoanalysis, 1940.

Chapter 8. Archaic Mammals and Hearing and Craik's Internal Map World

...high-pitched squeals of the mammals: Smith, J.C. and Sales, G.D., "Ultrasonic behavior and mother-infant interactions in rodents". In Bell, R.W. and Sotherman, W.P. (eds.), Infant Crying: Theoretical and Research Perspectives, SP Medical and Scientific Books, New York, 1985, pp.307-323.

...triune brain: MacLean, P.D., "The triune brain: emotion and scientific bias". In Schmidt, FO (ed), The Neurosciences Second Study Program, Rockefeller University Press, New York, 1970.

...separation call: MacLean, P.D., "Brain evolution relating to family, play, and the separation call", Arch. Gen. Psychiatry 42, 405-417, 1985.

...junkyard in the dark: McLoughlin, J.C., Synapsida, Viking Press, 1980.

...upper cortex...recurrent axons were excitatory: McGuire, B.A., J. Comp. Neurol. 305, 370-392, 1991. Quoted in Calvin, W.H., The Cerebral Code, MIT Press, 1998.

...spatial map arranged frequency......auditory cortex map: King, A.J. and Moore, D.R., "Plasticity of auditory maps in the brain", Trends in Neuroscience 14, 31-36, 1991.

...Fourier transform: Fourier first published in 1822, and his work is in any book on signals analysis such as Papoulis, A., Probability, Random

Variables and Stochastic Processes, McGraw Hill, New York, 1965.

...engram: Lashley, K.S., "In search of the engram", Symposium of the Society of Experimental Biology Vol.4, 454-482, 1950.

...internal representation: Craik, K., Nature of Explanation, Cambridge Univ. Press, Cambridge, 1943. There were precursors to this idea. The physiologist, Heinrich Helmholtz, in the 1890's thought that the brain made "unconscious inferences" about sensory inputs. Emanuel Kant in Critique of Pure Reason in 1781 discussed ideas somewhat like this.

...mental model system: Johnson-Laird, P.N., Mental Models, Cambridge Univ. Press, Cambridge, 1983.

...generative perceptual system: Hinton, G.E., "Computation by neural networks", Nature Neuroscience 3 (Suppl.), 1170-1172, Nov. 2000; and Hinton, G.E. and Sejnowski, T.J., "Learning and re-learning in Boltzman machines" in Rumelhart, D.E. and McClelland, J.L.(eds), Parallel Distributed Processing: Explorations in the Microstructure of Cognition, The MIT Press, 1986.

...predictive coding: Rao, R.N.P. and Ballard, D.H.,, "Predictive coding in the visual cortex...", Nature Neuroscience 2, 79-87, 1999.

...dynamic linear system: Staff of Research and Education Association, Theory of Linear Systems, Research and Education Association, New York, 1982.

...perceptual recognition failure...sensing stripped of meaning... (Seelenblindheit or soul blindness): Lissauer, H., "Ein Fall von Seelenblindheit nebst einem Beitrage zur Theori der selben", Arch. Psychiatr. Nervenkrankh 21,222-270, 1890. Quoted in, Feinberg, T.E. and Farah, M.J,., Behavioural Neurology and Neuropsychiatry, McGraw-Hill, 1997.

...beep-boop paradigm...P300: Sutton, S. et al, "Evoked-potential correlates of stimulus uncertainty", Science 150, 1187-1188, 1965, and Hillyard, S.A. and Picton, T.H., "Event-related potentials and selective information processing in man" in Desmedt, J. (Ed.), Cerebral Potentials in man: The Brussels Symposium, Oxford Unive. Press, London, 1977, and Picton, T. ,"The P300 wave of the human event-related potential", J. Clin. Neurophysiology 9(4), 456-479, 1992.

...sensory processing streams: Pandy, D.N. and Seltzer, B., "Association areas of the cerebral cortex", Trends in Neuroscience, 386-390, November 1982.

..."reality is a dream": Llinas, R., " 'Mindedness' as a functional state of the brain". In Blakemore, C. and Greenfield, S. (eds), Mindwaves, Blackwell Books, Oxford, 1987, pp. 339-358.

Chapter 9. Modern Mammals and Vision and Perceptual Sense Organs

..."The action center of the central nervous system moved up into the brain.": Allman, J.M., Evolving Brains, Scientific American library, W. H. Freeman, New York, 1999.

...ascending reticular activating system: Von Economo did the early work with his paper on World War I encephalitis lethargica patients, but Moruzzi and Magoun investigated and described it in detail in 1949. See discussion in Posner, J.B. et al, Plum and Posner's Diagnosis of Stupor and Coma, fourth edition, Oxford Un. Press, 2007.

...center-surround and line detection: Hubel, D.H. and Weisel, T.N.," Functional architecture of the macaque monkey visual cortex", Proc. Royal Soc. Lond., B. Biol. Sci. 198, 1-59, 1977.

...three cells to oriented line detector: Illustration after one in Mountcastle V.B., Medical Physiology, thirteenth edition, Mosely Co., 1974.

...primal sketch: Marr, D., Vision, MIT Press, 1982.

..."cortex...similar structure": Barlow, H.B., "The cortex as a model builder", in Rose, D. and Dobson, V.G. (eds), Models of the Visual Cortex, Wiley and Sons, 1985.

...temporal cortex...remembering what you have seen: Mishkin, M., "A memory system in the monkey", Phil. Trans. R. Soc. London, B 298, 85-95, 1982.

...Linsker, R., "From basic network principles to neural architecture:

Emergence of orientation columns" Proc. Natl. Acad. Sci. 83,8779-8783, 1986.

… sutured…eye…shut: Hubel, D.H. and Weisel, T.H., "The period of susceptibility to the effect of unilateral eye closure", J. Physiol. 206(2), 419, 1970.

…sutured eye shut…narrow cortical processing columns: LeVay, S., Wiesel, T.N., Hubel, D.H., "The development of ocular dominance columns in normal and visually deprived monkeys", J Comp Neurol. 191(1):1-51, 1980.

… rooms with no horizontal lines: Blakemore, C. and Cooper, G.F., "Development of the brain depends on the visual environment", Nature 228, 477-478, 1970.

…phantom triangle: Originally in Kanizsa, G., Rivista di Psicologia 49(1), 7-30, 1955.

…Necker cube illusion…Rubin vase-face illusion: L.A. Necker published in the London and Edinburgh Magazine and Science Journal 1(5), 329, 1832. The vase first appeared in E.J. Rubin's psychology thesis in 1915.

…80% cortical input to visual cortex: Gregory, R.L., "Knowledge in perception and illusion", Phil. Trans. R. Soc. Lond. B 352, 1121-1128, 1997.

…perceptions as hypotheses: Gregory said, "…predictive hypotheses of the external world and of ourselves… [which are] our most immediate reality" in Gregory, R.L., "Perceptions as hypotheses", Phil. Trans.R. Soc. Lond. B 290,181-197, 1980. He notes Heinrich Helmholz's prior idea in 1890 of the brain making "unconscious inferences" about sensory stimuli.

…perceptual vegetable recognition: Humphreys, G.M. and Riddoch, M.J., "On telling your fruit from your vegetables: a consideration of category-specific deficits after brain damage", Trends in Neuroscience 10 (4), 145-148, 1987.

…perceptual facial recognition…prosopagnosia: Damsio, A.R. et al,

"Prosopagnosia: anatomical basis and behavioural mechanisms", Neurology 32, 331-341, 1982.

...angular gyrus syndrome...Einstein's brain: Witelson, S.F. et al, "The exceptional brain of Albert Einstein", Lancet 353, 2149-2153, 1999.

...migraine scintillating scotomata...electrical disturbance: First described as "spreading depression of the activity of the cerebral cortex" by A.P. Leao in 1944, quoted in and its relation to migraine aura summarized in Jankovic, J. et al (eds), Bradley and Daroff 's Neurology in Clinical Practice, Elsevier, 2021.

Chapter 10. Hominids and Hands and Behavioral Motor Organs and Lorenz' World

...cat on treadmill: Kandel, E.R. and Schwartz, J.H., Principles of Neural Science, Third edition, Elsevier, NY, 1991.

...Gait ignition failure...gait apraxia: Adams, R.D. and Victor, M.V., Principles of Neurology, Fifth edition, McGraw Hill, NY, 1993.

...Libet, B.," Unconscious cerebral initiative and the role of conscious will in voluntary action", Behavioral and Brain Sciences, 8(4), 529–566, 1985.

...sensory homunculus: Penfield, W. and Rasmussen, T., The Cerebral Cortex of Man, MacMillan, NY, 1950.

...Ritter self-organizing map experiment: Ritter, H., "Large scale simulation of a self-organizing neural network: Formation of a somatotopic map", Proc. Int. Conf. on Parallel Processing in Neural Systems and Computers, Elsevier 1990.

...amputating monkey digits: Merzenich, M.M. et al, J. Comp. Neurol. 224,591, 1984. Also, Cox, J.L., "The brain's dynamic way of keeping in touch", Science 225, 820-821, 1984.

...cushion and movement internal representations: Gordon, A.M. and Forssberg, H., "Development of neural mechanisms underlying grasping in children". In Connolly, K.J. and Forssberg, H., (eds), Neurophysiology and Neuropsychology of Motor Development", MacKeith Press, London, 1997. Also, Goodwin, A.W., "Sensorimotor coordination in cerebral palsy", Lancet 353, 2090-2091, 1999.

...localization of man-made objects: Chao, L.L. and Martin, A,.., "Representation of Manipulable Man-Made Objects in the Dorsal Stream", NeuroImage 12(4), 478-484, 2000.

...motor behavior organs: Lorenz, K., "Uber die Bildung de Instiktbegriffes", Die Naturwissenschaften 25, 289-300, 1982.

...Piaget: Piaget, J., The Child and Reality, Penguin Books, Harmondsworth, UK, 1976.

Chapter 11. Homo Sapiens and Directed Attention and Consciousness and Popper's World

...frontal lobe myelination: Schoenemann, P.T., Sheehan, M.J., and Glotzer, L.D., "Prefrontal white matter volume is disproportionately larger in humans than other primates", Nature Neuroscience 8(2), 242-252, 2005.

...frontal lobes and delay tasks...Wisconsin card sort..." not the same Gage" ...working memory...executive function: Standard textbook material. See, for example, Stuss, D.T. and Benson, D.F., The Frontal Lobes, Raven, New York, 1986.

... ADD and ADHD: Mink, J. W., "Faulty Brakes: Inhibitory processes in attention-deficit/hyperactivity disorder", Neurology 76,592-3, 2011.

...the P300 evoked-potential and novelty: Daffner, K,.R., "The frontal and executive systems". In Samuels, M,.A. (ed), Intensive Review of Neurology, Harvard Medical School, 29 Sept. – 3 Oct., 2003.

...multi-modal representations: Pandya, D.N. and Seltzer, B., "Association areas of the cerebral cortex", Trends in Neurosci, 386-390, (Nov.) 1982.

...synesthesia: Cytowic, R.E. and Wood, F., "Synesthesia 1. A review of the major theories and their brain basis", Brain and Cognition 1, 23-35, 1982.

..."kiki" and "bouba": Ramachandran, V.S. and Hubbard, E.M.,"

Hearing colors, tasting shapes", Scientific America 53-59, (May) 2003.

...Aristotle...only the common sense went on to act on the rest of the nervous system.: In Harth, E., Windows of the Mind, William and Morrow, New York, 1982.

...nucleus reticularis and sensory gating: Skinner, J.E. and Yingling, C.D., "Central gating mechanisms that regulate event-related potentials and behavior: a neural model for attention". In Desmedt, J.E. (Ed.), Progress in Clinical Neurophysiology: Attention, Voluntary Contraction and Event-related Potentials, Vol.1, Karger, Basel, 1977, pp. 30-69, and 70-96, and also, Yingling, C.D. and Skinner, J.E., "Gating of thalamic input to cerebral cortex by nucleus reticularis thalami", Prog. Clin. Neurophysiol. 1, 70-96, 1977.

...sensory neglect: Mesulam, M.M., "A cortical network for directed attention and unilateral neglect", Ann. Neurol. 10, 309-325, 1981.

...N100 reflects the act of attending: Hillyard, S.A., Hink, R.F. et al, "Electrical signs of selective attention in the human brain", Science 182, 177-179, 1973.

...Functional Homunculus Brain Model: Haugland, J., "Semantic Engines: An Introduction to Mind Design" in Haugland, J. (ed.), Mind design: Philosophy, Psychology, Artificial Intelligence, MIT Press, Cambridge, 1981.

...searchlight of consciousness: Crick, F.C., "Functions of the thalamic reticular complex: the searchlight hypothesis", Proc. Nat. Acad. Sci. 81, 4586-90, 1984.

...claustrum: Crick, F.C. and Koch, C.,, "What is the function of the claustrum", Philosophical Transactions of the Royal Society Lond B Biol Sci 360(1458), 1271-9, 2005.

...electrode in claustrum: Koubeissi, M.Z,. et al, "Electrical stimulation of a small brain area reversibly disrupts consciousness", Epilepsy and behavior 37, 32-35, 2014.

...coma...awareness...persistent vegetative state: Young, G.B. and Pigott, S, "Neurobiological basis of consciousness", Arch. Neurol. 56, 153-157, 1999; and Posner, J,.B. et al, Plum and Posner's Diagnosis of

Stupor and Coma, fourth edition, Oxford Univ. Press, 2007.

...centipede in a ditch: The Centipede's Dilemma. Attributed to Craster, K., Pinafore Poems, 1874.

...sweaty T-shirt studies: Wedekind, C., "MHC-dependent preferences in humans", Proc. R. Soc. Lond. 206(1355), 245-249, 1995.

..." generate-and-test levels": Dennett, D.C., Darwin's Dangerous Idea: Evolution and the Meanings of Life, Simon and Schuster, New York, 1995.

... "replicating autonomous toolkits": Dobson, V.G., "Models and metaphysics: the nature of explanation revisited". In Rose, D. and Dobson, V.G. (eds.), Models of the Visual Cortex, J. Wiley and Sons Ltd, 1985.

..." permits our hypotheses to die in our stead": Popper, K. in Popper, K. and Eccles, J., The Self and Its Brain, Springer-Verlag, London, 1977, p 210.

Chapter 12. Homo Concept User and Conceptual Sensory Organs and Democritas' World

...concept as a sense organ: Erich Harth attributes this concept to Democritas in his Windows of the Mind, William and Morrow, New York, 1982. Since Democritas left us only fragments it is hard to be sure if this is exactly what he meant, but he did consider perception to be flawed and did refer to reason as the "genuine organ of knowledge". In Magill, F. (ed), World Philosophy Vol.1, Salem Press, New Jersey, 1983.

...conceptual agnosia: The term is a grammatical sin. The combining of the Latin, conceptus, and the Greek, gnosis, is an error so grave that Fowler calls it a barbarism. See Gowers, E. (ed), Fowler's Modern English Usage, Second Edition, Oxford Univ. Press, 1968. The classical Greek word for thought is noos, and the proper term, as far as I can work it out, would be anoognosia, which looks and sounds awful.

...Plato and quality: Barfield, O., History in English Words, Faber and Faber, London, 1953.

...Richard Gregory...a concept is a mind tool...: Dennett credits

Gregory with this concept and particularly of words being "mind tools". Dennett, D.C., Darwin's Dangerous Idea, Simon and Shuster, New York, 1995.

..." theoretical propensities": Miller, D., Popper Selections, Princeton Un. Press, Princeton, 1985.

...characteristic functions: These are also called eigenfunctions and pass through the matrix unchanged or changed only in size. They could be thought of as preferred inputs.

...Bayesian brain hypothesis: Dayan, P,., Hinton, G.E., Neal, R.M., "The Helmholtz machine", Neural Computation, 7, 889–904, 1995.They credit the early physiologist, Helmholtz, in the mid nineteenth century with the initial idea, but Hinton's group developed it as a neural network concept.

...Bayes Rule: The rule says that, if H means I see a horse and HF means I see hoofprints, then the probability of both at once is,

prob(H and HF) = prob(H /given HF). prob (HF)

= prob (HF/given H). prob(H)

and the probability that I see a horse, given that I see hoofprints, is

prob(H /given HF) = prob(H and HF)/ prob (HF)

= prob (HF/given H). prob(H) / prob (HF)

and, in Africa, prob(H) is much smaller and prob(zebra) is much larger.

...causality: Karl Friston has suggested that if the brain is making inferences about the causes of its sensation, then it must have a model of the causal relationship, in Friston, K., "The free-energy principle: a rough guide to the brain?", Trends in Cognitive Sciences 13 (7) 293-300, 2009.

...Jean Piaget: Inhelder, B. and Piaget, J., The Growth of Logical Thinking from Childhood to Adolescence, Basic Books, New York, 1958.

...two-kitten experiment: Held, R. and Hein, A.," Movement-produced stimulation in the development of visually guided behavior", J. Comp. and Physiol. Psychol. 56,872-876, 1963.

...Ashby, W.R., Design for a Brain, Chapman and Hall, London, 1952.

...Alice..." just a bag of neurons": Crick, F.C., The Amazing Hypothesis, Scribners, NY, 1994.

...rubber hand illusion: Botvinick, M. and Cohen, J., "Rubber hands 'feel' touch that eyes see", Nature 391, 756, 1998.

...hemisphere disconnection syndrome: Sperry, R.W., "Hemisphere disconnection and unity in conscious awareness", Am. Psychol. 23, 723-733, 1968.

...alien hand syndrome: Banks, G. et al, "The alien hand syndrome: clinical and postmortem findings", Arch. Neurol.46, 456-459, 1989.

...frontal lobe personality change: Miller, B.L. et al, "Neuroanatomy of the self", Neurology 57, 817-821, 2001.

...learn to be selves: Popper, K., "The Self (1977)", in Miller, D., Popper Selections, Princeton University Press, Princeton, N.J., 1985.

...hermeneutics: Ricoeur, J. P.G., Freud and Philosophy: An Essay in Interpretation, Yale Univ. Press, New Haven, 1970.

...theory of other minds: Premack, D. and Woodruff, G., Behav. Brain Sci., 512-526,1978.

...autism: Frith, U., Morton, J., and Leslie, A., " The cognitive basis of a biological disorder: autism", Trends in Neuroscience 14, 433-438, 1991.

...anthropologist on Mars: Sacks, O., An Anthropologist on Mars: Seven Paradoxical tales, Vintage Books, NY, 1995.

...red dot experiment: Gallop, G.G. Jr., "Chimpanzees: self-recognition", Science 167, 86-87, 1970.

... "understand" that object, which is to say we understand how it works: Johnson-Laird, P.N., Mental Models, Harvard Univ. Press, Cambridge, 1983, and Johnson-Laird, P.N., "A history of mental models", in Manktelow, K., and Chung, M.C. (eds), Psychology of Reasoning, Psychol Press, 2004.

...Turing test: Turing, A.M.," Computing machinery and intelligence", Mind 59 (236), 4-30, 1950.

...self-programming computer: First reported in von Neumann, J., "First draft of a report on EDVAC", Contract No. W-670-ORD-4926, Moore School of Electrical Engineering, Univ. of Pennsylvania, Philadelphia, PA, 30 June 1945.

...novelty...neural network errors: Cooper, L.N., "A possible organization of animal learning and memory", Nobel 24, 252-264. In Lundquist, B. and Lundquist, A. (eds), Proceedings of the Nobel Symposium on the Collective Properties of Physical Systems, Acad. Press, NY, 1972.

...Marvin Gardner quoted Piet Hein: In Gardner, M., "Mathematical games: free will revisited, with a mind-bending prediction paradox by William Newcomb", Scientific American 229, 104-9, 1973.

...testable hypotheses: Popper demarcates science from metaphysics and pseudo-science by hypotheses that can be tested and proved wrong. See Popper, K., "The problem of demarcation(1974)", Miller, D. , Popper Selections, Princeton University Press, NJ, 1985.

... Hegel's perfect apprehension: Richeimer, J., "Hegel", in Shutt, T.B., Odyssey of the West, Volume V, The Modern Scholar, Recorded Books, Prince Frederich, MD, 2008.

Chapter 13. Homo Language User and Semantic Tools and Gregory's World

...chiseled images of the phases of the moon: Eccles, J.C., Evolution of the Brain: Creation of the Self, Routledge, London, 1989.

...descriptive speech...vervet monkeys: Buhler, K., Quoted in Popper, K., "The mind-body problem (1977)", in Miller, D., Popper Selections, Princeton University Press, Princeton, N.J., 1985.

...speech recognition center...speech production center: See Victor and Adams.

...speech comprehension centers...Lichtstein's house: Ludwig Lichtstein described what he called "Uber aphasie" in 1885. All three aphasias are discussed in Butterworth, B., "Chapter 21: Aphasia and models of language production and perception" in Blanken, G., Dittman, J. et al (eds), Linguistic Disorders and Pathologies, de Gruyer, Berlin, 1993.

...isolation of speech circuit: Shim, H. and Grabowski, T.J., "Comprehension", in Behavioral Neurology, Continuum, Vol. 16, No. 4. , American Academy of Neurology, Aug. 2010.

...syntax: Chomsky, N., Aspects of the Theory of Syntax, MIT Press, Cambridge, 1965.

...speech maps: Ritter, H. and Kohonen, T., "Self-Organizing semantic maps", Bio. Cybernetics 61, 241-254, 1989.

..." blundered into language, and then became smart ... primarily a representation system": Bickerton, D., Language and Human Behavior, Univ. Washington Press, Seattle, 1995.

..."verbs of a feather...", Gleitman, L.R., Liberman, M.Y., McLenore, C.A., Partee, B.H., "The impossibility of language acquisition (and how they do it)", Annual Review of Linguistics 5, 1-24, 2019.

...triangles' gods: Baron de Montesquieu wondered if triangles' gods would have three sides.

..." Experience is a vast web...": Henry, J., The Art of Fiction, 1884.

...multi-modal sensory areas gave rise to the speech system: Geschwind, N., "The development of the brain and the evolution of language", Monograph Series in Language and Linguistics, 1, 155-169, 1964.

...situation labels: Hoffstadter, D., "Analogy as the core of cognition" in: Gentner, D., Holyoak, K.J., and Kokinov, B.N. (eds.), The Analogical Mind: Perspectives from Cognitive Science, The MIT Press/Bradford Book, Cambridge MA, 2001.

...metaphoric thinking: Lakoff, G. and Johnson, M., Metaphors We Live By, Univ. Chicago Press, Chicago, 1980.

...to breathe...to be: Barfield, O., History in English Words, Faber and Faber, London, 1953.

... "spin complex chains of hypothetical cause and effect": Dennett credits Gregory with the concept of Gregorian creatures spinning such chains in Dennett, D.C., Darwin's Dangerous Idea, Simon and Shuster, New York, 1995.

Chapter 14. Homo Abstraction User and Memes and Plato's World

...abstraction...Plato's Forms: Barfield, O., History in English Words, Faber and Faber, London, 1953.

...World III: Popper, K., "The mind-body problem (1977)". In Miller, D. (ed), Popper Selections, Princeton University Press, NJ, 1985.

...Xenophanes: From Aristotle's discussion of Xenophanes quoted in the prologue of Hildebrandt, S. and Tromba, A., Mathematics and Optimal Form, Scientific American Books, WH Freeman and Co., New York, 1985.

...thinking as pattern recognition: Margolis, H., Patterns, Thinking and Cognition, Univ. Chicago Press, 1988.

...analogy...core of thinking: Hoffstadter, D., "Analogy as the core of cognition" in Gentner, D., Holyoak, K.J., and Kokinov, B.N. (eds.), The Analogical Mind: Perspectives from Cognitive Science, The MIT Press/Bradford Book, Cambridge MA, 2001.

...Dostoyevsky: Alajouanine, T., "Dostoiewski's epilepsy", Brain 86, 209-218, 1063, quoted in Rolak, L.A., Neurology Secrets, Hanley and Belfus, 1993.

...stroke...Nirvana: Taylor, J.B., My Stroke of Insight, Viking Press, 2008.

...memes: Dawkins, R., The Selfish Gene, Oxford Univ. Press, Oxford, 1976.

...ideas evolving: Huxley, T.H., "The coming of age of 'The origin of species'", Science 1(2), 15–20, 1880.

...superorganism: The concept originated with Herbert Spenser and was first called, "Super-organic evolution" in his Principles of Sociology in 1876. The correct term is said to be supraorganism, but most use the other.

..." but the dreams their children dreamed...": Mary Elizabeth Coleridge, "Vale - Egypt's Might Is Tumbled Down," 1886.

...nervous system also models the world: Dobson, V.G., "Models and metaphysics: the nature of explanation revisited". In Rose, D. and Dobson, V.G. (Eds), Models of the Visual Cortex, Wiley and Sons Ltd, 1985.

..." fin...water...The whole organism can be considered a coded representation of its environment." Young, J.Z., Programs of the Brain, Oxford Univ. Press, Oxford, 1978.

About the Author

Michael Blake Evans, B.Eng. (Electrical Engineering), M.Sc. (Electrical Engineering), M.D., Specialist Certificate (Psychiatry}, Fellow of the Royal College of Physicians and Surgeons of Canada , Diplomate of American Board of Psychiatry and Neurology (Neurology), American Academy of Neurology; Captain (Ret.), Royal Canadian Corps of Signals.

The author's medical and engineering education is the basis of this book. He studied electrical engineering as an undergraduate and went on to do graduate work on signals processing systems. He studied medicine, qualified in neurology and psychiatry, and did a research fellowship on cerebral evoked-potentials.

He worked and learned as a private practice neurologist, as an assistant professor of neurology, and as a practitioner of clinical neurophysiology or interpreter of electrical measurements of muscle, nerve, and central nervous system including evoked-potentials, at first on paper and later over the internet.

He would like to acknowledge the contributions to his education and thank Professor George Atkin, Ph. D, who supervised investigations of the superesolution problem in radio astronomy and interested him in distributed representations of information and their signals processing, and Professor James E. Skinner, Ph.D., who supervised investigations of cerebral evoked-potentials and interested him in the olfactory and directed-attention systems.

Those who have contributed in universities, engineering departments, and clinical medicine departments are too many to list and thank, but he would like to acknowledge the lessons learned in the three universities he attended, College Militaire Royale Saint-Jean in Saint-Jean-sur-Richelieu, Quebec, Royal Military College in Kingston, Ontario, and Queen's University, also in Kingston; in the clinical departments and corridors of the teaching hospitals he trained in, particularly Kingston General Hospital, Vancouver General Hospital, and the seven Baylor College of Medicine hospitals of the Texas Medical Center in Houston; and in the hospitals and corridors walked since residency.